CARL FRIEDRICH GAUSS

CARL FRIEDRICH GAUSS
A Biography by Tord Hall
Translated by Albert Froderberg

THE MIT PRESS
Cambridge, Massachusetts, and London, England

Originally published in Swedish by
Bokförlaget Prisma, Stockholm,
under the title "Gauss, Matematikernas konung"

English translation Copyright © 1970 by
The Massachusetts Institute of Technology

Designed by Dwight E. Agner

Set in Monophoto Times Roman
Printed and bound in the United States of America

ISBN 0 262 08040 0 (hardcover)
ISBN 0 262 58017 9 (paperback)

Library of Congress catalog card number: 71-110227

TRANSLATOR'S FOREWORD

I thank Universitetslektor Tord Hall for the help he gave me in preparing this translation of his book. I have attempted to reproduce the character of the original Swedish edition as closely as possible. Some slight changes and corrections, mostly supplied by the author, are made in this edition. Thanks are also due to Professors Neil R. Gray, Edwin Hewitt, and Kenneth O. May. Each of them read all or part of the translation and offered valuable advice.

Bellingham, Washington   Albert Froderberg
March 1970

PREFACE

Just as the lion in the rhetoric of an older time was called king of beasts, Carl Friedrich Gauss was called the king of mathematicians, *Mathematicorum princeps*.
He was fully mature at the time of his debut as an eighteen-year-old, and for more than fifty years he maintained the same high standard. He worked with equal ease in all areas of pure and applied mathematics: in number theory, algebra, function theory, differential geometry, probability theory, astronomy, mechanics, geodesy, hydrostatics, electrostatics, magnetism, optics, and so on.
Because of the universal range of his work, he became the foremost representative of the scientific viewpoint in the nineteenth century, and through his research on curved surfaces he laid the mathematical groundwork for the twentieth-century viewpoint that is based on the general theory of relativity.
Usually only Archimedes and Newton are mentioned as his equals. While Archimedes' and Newton's work have been part of school courses for some time and have more or less become common knowledge, Gauss's mathematics lies well beyond the high-school curriculum —even well beyond introductory university courses. The same is true of most of his contributions in applied mathematics.
In spite of this it is not impossible, by using relatively simple devices, to give a wider circle of readers some notion of the different fundamental ideas in Gauss's work. His research on prime numbers and his notion of congruence, fundamental in his "Arithmetical Investigations," demand no special prior knowledge, for example, and the same is true of his general ideas about non-Euclidean geometry—for the last named subject it is even preferable to have no prior notions at

all. But in several other instances there do remain insurmountable obstacles to a popular presentation.

One way out would be to write a book about Gauss without going into his mathematics, but I think that to do this would be as meaningless as writing a book about Beethoven without going into his music. Therefore, I have not avoided mathematical technicalities, but I have tried to keep them to a minimum. An exact presentation of proofs and the like is obviously beyond the scope of this book. I have tried to give Gauss's most important results by formulating the problems, saying something about their origin, and illustrating them with concrete examples. I have also tried to illustrate Gauss's method of working and the demands he placed upon himself.

Nonetheless, our mathematical selections must be a subjective choice, a selection from the general contents of Gauss's collected work, the publication of which occupied ten experts in pure and applied mathematics during the seventy years from 1863 to 1933.

Uppsala, January 1965   *Tord Hall*

CONTENTS

CARL FRIEDRICH GAUSS

ONE ANCESTRY AND ENVIRONMENT

Neither in Gauss's descent nor in his childhood environment is there any certain indication of his coming life's work. On his father's side there appear owners of small farms, farm workers, and laborers in Braunschweig, which is now a part of West Germany. They worked hard for their living and could not have had much time left over for other activities. On his mother's side the economic and social positions were often better. There we find pastors, well-to-do farmers, masons, and holders of honorable secular or ecclesiastic posts. Gauss's talents therefore are often said to be inherited through his mother. But pastors, masons, and church wardens can offer no definite explanation for the appearance of such a genius, even though Gauss's father was quite a good calculator. We must certainly acknowledge our uncertainty on this point, and if we say that his talents were the consequence of a gene mutation caused by cosmic radiation, then we have only admitted about the same thing.

The family name had different spellings in the early days. His paternal grandfather, Jürgen Goos, settled down in the city of Braunschweig, then the capital of the Duchy of Braunschweig, in 1739. His father, Gebhard Dietrich Gauss, was born in 1744. He was a jack-of-all-trades. Through hard work as a stone mason, canal worker, and gardener, among other occupations, he at last became full owner of the house at Wilhelmstrasse 30 which had been purchased by Jürgen Goos in 1753—with a heavy mortgage. Since Gebhard calculated and wrote well, he was entrusted with the office of treasurer of a burial fund.

His first wife died in 1775, and the following year he married Dorothea Benze, who was born in 1743. The only child of this new union was Johann Friedrich Carl, who was born on April 30, 1777, in the house on Wilhelmstrasse (which later became a museum and was destroyed in a bombing raid during the Second World War).

Gauss's maternal grandfather, Kristoffer Benze, was a mason in the village of Velpke outside Braunschweig. Because he worked in sandstone his lungs were affected, and he died when he was only thirty years old. Dorothea's younger brother, Johann Friedrich, was both gifted and original. He taught himself to be a fine damask weaver. When he died in 1809, Gauss declared that the world had lost a genius. On this point we have only Gauss's evidence to follow, and because he was so bound to his mother in his younger days, his outlook may be considered subjective. In any case we can observe, in present-day terminology, that the uncle—as opposed to Gauss himself—belonged to the unexploited reserve of the gifted.

Dorothea never learned to write, and she could scarcely read. But she had a fine natural intellect, good humor, and strong character. Carl Friedrich was always her dominating interest. She spent the last twenty-two years of her life with Gauss at the observatory in Göttingen. She was ninety-seven years old when she died in the year 1839. Gebhard Gauss died in Braunschweig in 1808, and even though he was economically independent toward the end of his life, one cannot say that he ever became wealthy.

In 1810 Gauss described his parents in a letter to Minna Waldeck, who later became his second wife: "My father was completely honest, in many ways worthy of respect, and certainly a well-regarded man; but in his home he was tyrannical, coarse, and violent . . . he never had my full confidence when I was a child, but no real break ever resulted, since I became independent of him very early. . . .

My mother was born fifty kilometers from Braunschweig, and she worked there several years as a servant girl. She married my father in 1776, and has no children other than myself. Her marriage was not happy, but that for the most part was due to outside circumstances and the fact that the two personalities were not compatible. My mother is certainly a very good woman, who is not unworthy of her child's love."

TWO CHILDHOOD

The Mathematical Prodigy.

Biographies about or by great men generally contain more or less note-worthy anecdotes, intended to illustrate the budding genius. It is a field in which memory gladly accommodates itself to a fixed path and where imagination easily overtakes the uncertain facts. The situation is especially pernicious in the case of child prodigies, who are often encountered in mathematics, music, and chess. Myths appear with treacherous ease.

Gauss was a mathematical prodigy—it is certain that he was one of the most outstanding examples of this genre, but basically this is unimportant. First-hand accounts of this come from Gauss himself, who in his old age liked to talk of his childhood. From a critical viewpoint they are naturally suspect, but his stories have been confirmed by other persons, and in any case they have anecdotal interest.

During the summers Gebhard Gauss was foreman for a masonry firm, and on Saturdays he used to pay the week's wages to his workers. One time, just as Gebhard was about to pay a sum, Carl Friedrich rose up and said: "Papa, you have made a mistake," and then he named another figure. The three-year-old child had followed the calculation from the floor, and to the open-mouthed surprise of those standing around, a check showed that Carl Friedrich was correct.

Gauss used to say laughingly that he could reckon before he could talk. He asked the adults how to pronounce the letters of the alphabet and learned to read by himself.

Schooling.

When Carl Friedrich was seven years old he enrolled in St. Catherine elementary school. His teacher was J. G. Büttner. The large classroom had a low ceiling, and the schoolmaster walked about on the uneven floor, cane in hand, among his approximately one hundred pupils. Caning was the foremost pedagogical aid both for learning and discipline, and Büttner is thought to have used it constantly, either as a consequence of necessity or because of his temperament. Gauss stayed in these surroundings for two years without any ill consequences. It is the traditional picture of that period's public education, when the caning pedagogy was generally accepted—by the adults of course—but we shall soon see that Büttner was more likely above than below average among his colleagues.

When Gauss was about ten years old and was attending the arithmetic class, Büttner asked the following twister of his pupils: "Write down all the whole numbers from 1 to 100 and add up their sum." When the class had a task of that sort they would do the following: the first to finish would go forward to the schoolmaster's desk with his slate and put it down, the next who finished would place his slate upon the first, and so on in a growing pile. The problem is not difficult for a person familiar with arithmetic progressions, but the boys were still at the beginner's level, and Büttner certainly thought that he would be able to take it easy for a good while. But he thought wrong. In a few seconds, Gauss laid his slate on the table, and at the same time he said in his Braunschweig dialect: "Ligget se" (there it lies). While the other pupils added until their brows began to sweat, Gauss sat calm and still, undisturbed by Büttner's scornful or suspicious glances.

At the end of the period the results were examined. Most of them were wrong and were corrected with the rattan cane. On Gauss's slate, which lay on the bottom, there was only one number: 5050. (It seems unnecessary to point out that this was correct.) Now Gauss had to explain to the amazed Büttner how he had found his result: $1 + 100 = 101$, $2 + 99 = 101$, $3 + 98 = 101$, and so on, until finally $49 + 52 = 101$ and $50 + 51 = 101$. This is a total of 50 pairs of numbers, each of which adds up to 101. Therefore, the whole sum is $50 \times 101 = 5050$. Thus Gauss had found the symmetry property of arithmetic progressions by pairing together the terms as one does when deriving the summation formula for an arbitrary arithmetic progression—a formula which

Gauss probably discovered on his own. What this actually entails is that one writes the series both "forward" and "backward"; that is

$$1 + 2 + \ldots + 99 + 100$$
$$100 + 99 + \ldots + 2 + 1.$$

Addition in the vertical columns gives 100 terms, each of which is equal to 101. Since this is twice the sum wanted, the answer is $50 \times 101 = 5050$.

The event is symbolic. For the rest of his life Gauss was to present his results in the same calm, matter-of-fact way, fully conscious of their correctness. The evidence of his struggles would be wiped away from the completed work in the same way. And, like Büttner, many learned persons would wish to be given a detailed explanation, but here a difference would appear, for Gauss would not feel compelled to give one.

This event also marks a turning point in Gauss's life. Büttner realized immediately that he could teach almost nothing to Gauss, and gave him a better arithmetic textbook especially ordered from Hamburg. Gauss also came into closer contact with the eighteen-year-old Martin Bartels, who assisted Büttner with his teaching. This was a lucky stroke, for Bartels, who later became a professor of mathematics, could understand and help Gauss far better than Büttner.

Either Büttner or Bartels, or both, visited Gauss's father to talk about the boy's education. Gebhard was accustomed to having his will regarded as law in the family. He had thought that his two sons should follow in his own footsteps—and Carl Friedrich's half-brother George, who was born of the father's first marriage, did so. Gebhard was reluctant and wondered—with perfectly good reason—how he could get hold of enough money for a higher education. Bartels and Büttner answered with the only argument that was customary, and often the only one possible, in those days: "We shall no doubt be able to find some distinguished person as protector or patron for such a genius." The result was a compromise. Gebhard permitted the boy to give up his usual evening chore of spinning a certain amount of flax. The spinning wheel disappeared—Gebhard is said to have made firewood of it—and in its place there appeared books. Gauss and Bartels used to sit and discuss mathematical problems late into the evening, and soon Bartels too had nothing further to teach Gauss.

In 1788 Gauss enrolled—almost against his father's will, according to his own statement—at the Gymnasium Catharineum in Braunschweig. Professor Hellwig, who was the mathematician there, returned Gauss's first written work with the comment that it was not necessary for such a gifted student to continue attending the lectures in his class. With the help of Bartels and the philologist Meyerhoff, Gauss soon surpassed his fellow students in classical languages as well.

Duke Ferdinand Becomes His Patron.

Through Bartels, Gauss came in contact with Professor Zimmermann at the Collegium Carolinum (which later became a technical institute), who in turn became the intermediary link between Gauss and his patron, the reigning Duke Carl Wilhelm Ferdinand of Braunschweig. Gauss was summoned to an audience by a court official and all went well. The duke undertook to supply whatever means should be needed "for the continued training of such a gifted person," beginning in 1791, and through his tact he evidently won the trust and devotion of the shy fourteen-year-old at once.

After the economic problems were solved in this feudal manner, it may be supposed that Gauss's father had nothing more against his son's continued studies. Gauss attended the Collegium Carolinum during the years 1792–1795. When he published his first scientific notice in 1796 (about the regular 17-sided polygon), Zimmermann introduced him to the circle of readers with a few lines; among other things he wrote that Gauss "here at Braunschweig has devoted himself to philosophy and literature with as much success as in higher mathematics."

First Independent Research. The Prime Numbers and Gauss's Hypothesis Concerning Their Distribution.

Before a maturing mathematician can make any contribution to his science, he must first master the results that are already known. He can do so partly by using a textbook or a teacher, and partly by thinking on his own. In his childhood surroundings Gauss could utilize the first of these possibilities only within narrow limits, and, considering his own great gifts, it is only natural that he used the second for the most part. In so doing he often rediscovered theorems that were already known, for example the formula for the sum of an arithmetic progression.

Like all other children, Gauss first encountered in his mathematical studies the *natural numbers* 1, 2, 3, 4, ... They look simple, but they conceal many of the hardest of mathematical problems. Among them, for example, are many questions about prime numbers. Before we go into Gauss's contributions to this field we shall state some definitions and known facts.

A *prime number* is a natural number that is larger than 1 and has no factors other than 1 and itself. Other natural numbers, if they are larger than 1, are called *composite numbers*. The number 1 belongs to neither of these classes; it stands alone. The number 2 is the only even prime number. The numbers 13 and 19 are prime numbers, while $6 = 2 \times 3$ and $25 = 5 \times 5$ are composite numbers. If we ignore the order of the factors, every composite number can be written in only one way as the product of prime numbers, which are called its *prime factors*.

We can always determine whether or not an arbitrary natural number x is a prime number by dividing x by the numbers 2, 3, 4, . . . , $x - 1$. If none of these divisions is possible without a remainder, then x is clearly a prime number. But this method soon becomes terribly laborious.

We can simplify the work by using square roots. The positive square root of x is denoted by $\sqrt{x}$ and it is defined as the number such that $\sqrt{x} \times \sqrt{x} = x$. From this we see, for example, that $\sqrt{25} = 5$, since $5 \times 5 = 25$, and $\sqrt{36} = 6$, since $6 \times 6 = 36$. In general $\sqrt{x}$ is not a natural number; for example, the square roots of the numbers from 26 to 35 all lie between 5 and 6, as can be seen from the foregoing examples.

A prime number clearly cannot have two prime factors that are both larger than its square root. This implies that each composite number has at least one prime factor that is less than or equal to its square root.

From this fact it follows in turn that if we know beforehand all of the prime numbers p that are less than or equal to $\sqrt{x}$—in mathematical language we write: all $p \leq \sqrt{x}$—then the prime-number question can be settled with much less work. It now suffices to divide x by each of the numbers p. The fact that x is a prime number is then equivalent to the statement that none of these divisions is possible without remainder.

Let us apply this to the number 61. The numbers $p \leq \sqrt{61}$ are 2, 3, 5, and 7. Since none of these numbers divides 61 without a remainder, 61 is a prime number. In the same way it can be shown that 143 is a composite number, since it is divisible by the last of the numbers 2, 3, 5, 7, and 11.

Knowledge of the prime numbers $p \leq \sqrt{x}$ can also be used to determine *all* of the prime numbers that are less than or equal to x. To illustrate this method with an example, we choose $x = 31$. The numbers $p \leq \sqrt{31}$ are 2, 3, and 5. We then write down the sequence of numbers 2–31 one after another according to the scheme below:

2	3	4	5		6	7	8	9	10	11
12	13	14	15	16	17	18	19	20	21
22	23	24	25	26	27	28	29	30	31

The vertical bar after the number 5 is to mark the largest prime number less than or equal to $\sqrt{31}$. Beginning with the number 2 we strike out every second number in the scheme. Beginning with the number 3 we strike out every third number, and finally beginning with 5 we strike out every fifth number. In this way every number to the right of the vertical bar in the scheme that is a multiple of any of the factors 2, 3, or 5 vanishes; that is to say, all of the numbers divisible by 2, 3, or 5 are eliminated. The remaining numbers that are caught by our "sieve" are

$$7, \quad 11, \quad 13, \quad 17, \quad 19, \quad 23, \quad 29, \quad 31.$$

If any number in this sequence were composite, it would have to contain at least one prime factor that is less than or equal to $\sqrt{31}$; that is, it would have to contain at least one of the numbers 2, 3, or 5 as a factor. But it was just those numbers (and no others) that were stricken from the list. Thus none of the numbers stuck in the sieve is composite. The only possibility remaining is that they are all prime numbers, and we see that we have found *all* of the prime numbers that are larger than $\sqrt{31}$ and less than or equal to 31.

The same procedure that we have used on the number 31 can be used on an arbitrary natural number x, as long as we know all of the primes $p \le \sqrt{x}$. For example, the reader can apply the method to $x = 100$, where the primes $p \le \sqrt{x}$ are 2, 3, 5, and 7. The result will be a table of primes for the numbers in the interval between 1 and 100. It is simplest to just strike out all multiples of 2, 3, 5, and 7 in the sequence 8 to 100; the remaining numbers are then the prime numbers between 10 and 100.

This simple and effective way of finding prime numbers was mentioned for the first time, as far as we know, by the Greek Eratosthenes, who was librarian at the museum in Alexandria, and who lived from 276 to 194 B.C. Because of this the method is usually called the *sieve of Eratosthenes*.

Although mathematics has made enormous progress since ancient times, there have been no revolutionary changes in Eratosthenes' technique. Even today, anyone wishing to produce a table of prime numbers on his own is still referred to Eratosthenes' sieve, in principle. When the number becomes quite large even this method becomes tremendously laborious, and in practice impossible to carry out even for the best calculator.

For certain numbers, for example those of the types $2^n + 1$ and $2^n - 1$, where n is a natural number, there exist arithmetical criteria that simplify the solution of the question of whether or not the number is a prime. In many other cases also, one may use results from number theory to simplify calculations that would otherwise be impossible even for today's best computers.

When one sieves the prime numbers, one of the first questions that comes to mind is the following: "Are there infinitely many prime numbers, or do they end with a particular number?" Euclid, who lived in Alexandria during the fourth century B.C., gave an elegant proof that the number of primes is infinite.

The next question is: "How is the set of prime numbers distributed among the natural numbers?" The Greeks could not answer that question, nor can we, except in an approximate way. An exact formula for the distribution of prime numbers—if it exists—is among the natural numbers' best hidden secrets.

If we knew such a formula we would be able to give an exact answer to this question: "How many prime numbers are there that are less than or equal to an arbitrary natural number x?" As things now stand, we can give only a very approximate answer.

The number of primes that are less than or equal to the natural number x

x	$\pi(x)$	x	$\pi(x)$
5	3	80	22
10	4	90	24
20	8	100	25
30	10	500	95
40	12	1,000	168
50	15	5,000	669
60	17	10,000	1,229
70	19	1,000,000	78,498

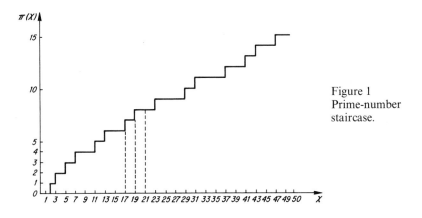

Figure 1
Prime-number
staircase.

Nevertheless it is certainly true that the number we seek is uniquely determined by x. In mathematical terminology we say that it is a *function of x*. This function is usually denoted by $\pi(x)$, which is defined as the number of primes that are less than or equal to x. We have, for example $\pi(10) = 4$, since the prime numbers equal to or less than 10 are 2, 3, 5, and 7; that is to say there are 4 of them; and $\pi(31) = 11$. (Cf. example on p. 9 in connection with the sieve of Eratosthenes.) The table opposite shows the value of $\pi(x)$ for a few selected values of x.

If we let x run through all the positive real numbers (not just the natural whole numbers), we see that the curve corresponding to $\pi(x)$ looks like a staircase, one that grows one step each time that x passes a prime number (Fig. 1). The different widths of the treads show that the prime numbers are irregularly distributed; this impression is confirmed if the curve is continued.

Since there are infinitely many prime numbers, $\pi(x)$ grows without bound as x does the same. From a graphical viewpoint this states that the staircase does not end with a high plateau, but instead grows in height toward infinity.

The investigation of the function $\pi(x)$ is the principal task in the theory of the distribution of prime numbers. Unfortunately one has had to be satisfied with attempts to find relatively simple functions that approximate $\pi(x)$ with as little error as possible, and it is here that Gauss finally comes into the story.

When he was about fifteen years old, that is, during 1792–1793, he investigated the distribution of prime numbers with the help of a table of prime numbers published by the Swiss Johann Lambert. Gauss divided the natural numbers into thousands, 1 to 1,000, 1,000 to 2,000, and so forth, and used Lambert's table to calculate the number of primes in each interval; that is to say, he determined $\pi(1,000)$, $\pi(2,000) - \pi(1,000)$, and so forth. In the following table, which only shows the beginning, these differences that measure the rate of growth of the prime numbers are denoted by $D(x)$. (Gauss also used other intervals, such as those of length 100 and 100,000.)

Distribution of prime numbers up to 10,000

x	$\pi(x)$	$D(x)$
1,000	168	168
2,000	303	135
3,000	430	127
4,000	550	120
5,000	669	119
6,000	783	114
7,000	900	117
8,000	1,007	107
9,000	1,117	110
10,000	1,229	112

The tendency of $D(x)$ to decrease slowly as x increases appears here, and this tendency remains as the table is extended. For the most part this growth continues at a slower and slower rate, without ever completely stopping; this implies that the prime numbers become generally sparser the farther out we go among the natural numbers. For the staircase of Fig. 1 the distance between steps will be longer and longer on the average, the farther out we go in the direction of the x-axis. (It is easy to see that there is no bound for the lengths of the intervals that contain no prime numbers.)

This general property of the prime numbers was already known, and since it is rather trivial it is of no importance whether or not Gauss discovered it by himself. But that which follows is not so trivial. By investigating different simple functions Gauss soon found that on the average $D(x)$ was inversely proportional to the natural logarithm of x, written log x.

Since he had found an expression for $D(x)$, he could determine an expression for $\pi(x)$ by a summation procedure. Summations are very closely related to integrals in mathematics, and the fifteen-year-old Gauss presented his hypothesis about the distribution of prime numbers in the form of an integral: *the number of primes that are less than or equal to the natural number x is given approximately by*

$$\int_2^x \frac{1}{\log n}\, dn.$$

Today the derivation of this formula lies at the university level of mathematics. Already we have an example of the numerical skill and analytical power that was so characteristic of Gauss.

Let us give a geometric interpretation of what this mathematical symbol means. In the growth function $D(x)$ we replace x by n in order to avoid confusing the notation. We will express the fact that $D(n)$ is inversely proportional to log n by the formula

$$D(n) \cong \frac{1}{\log n}.$$

The curve corresponding to $D(n)$ is sketched in Fig. 2, which shows only the part that corresponds to values of n in the interval 2 to 200. The geometrical interpretation is that the value of the integral

$$\int_2^x \frac{1}{\log n}\, dn$$

is equal to the area of the shaded region A in Fig. 2; that is to say, the area between the curve, the horizontal axis, and the vertical lines $n = 2$ and $n = x$. If the curve is drawn on graph paper, we can find an approximation to $\pi(x)$ by counting squares.

In the same way, the value of the integral

$$\int_{100}^{200} \frac{1}{\log n}\, dn$$

is equal to the area of the region B in Fig. 2, and this number is approximately equal to $\pi(200) - \pi(100)$; that is, the number of primes in the interval 100 to 200. Since the curve is almost horizontal in this region we can replace the area B by a rectangle with base 100 and height 0.2 units. Thus the area of B is approximately 20. The actual number of primes between 100 and 200 is 21.

In 1796, when the Austrian Georg von Vega published a table of prime numbers that went up to 400,031, Gauss could substantiate his hypothesis further. Later tables came out that went even farther and gave the same result. But Gauss was not satisfied with slavishly accepting the published figures. He used his ability for numerical calculation, which is unique in the history of mathematics, to check the tables by means of samples, among other things. He never bothered to draw up a systematic table of prime numbers on his own. It was the sort of unimaginative job that did not attract him, and in his time there were no computers (that surpass even Gauss many times over in speed of calculation). When he made his samples he certainly did not rely only on Eratosthenes' sieve but also used personal tricks of reckoning or number-theoretical theorems that he had discovered himself.

In a letter of 1849 to the astronomer Johann Franz Encke, Gauss tells about his youthful investigation of the distribution of prime numbers: ". . . and I have (since I lacked the patience to go through the whole series systematically) often used a spare quarter of an hour to investigate a thousand numbers here and there; at last I gave it up altogether, without ever finishing the first million."

As a result of his industry at this hobby he was able to inform Encke: "The thousand numbers between 101,000 and 102,000 bristles with errors in Lambert's supplement; in my copies I have crossed out

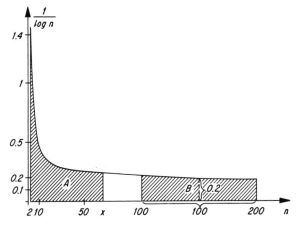

Figure 2
Graphic presentation of Gauss's hypothesis on the distribution of prime numbers.

seven numbers that are not primes, and in return put in two that were missing." In the same letter, after ! tells in a few lines how he found his integral formula for the approximation of $\pi(x)$, Gauss supplies a short extract from his table of the distribution of primes and the values of the integrals. It is reproduced here, with the addition that the values from the table of primes published in 1913 (reprinted in 1956) by the American Derrick N. Lehmer are put in the right-hand column for purposes of comparison:

Excerpt from Gauss's and Lehmer's tables for the distribution of prime numbers

x	$\pi(x)$ according to Gauss	$\int_2^x \frac{1}{\log n}\,dn$	Deviation $R(x)$	$\pi(x)$ according to Lehmer
500,000	41,556	41,606.4	+50.4	41,538
1,000,000	78,501	78,627.5	+126.5	78,498
1,500,000	114,112	114,263.1	+151.1	114,155
2,000,000	148,883	149,054.8	+171.8	148,933
2,500,000	183,016	183,245.0	+229.0	183,072
3,000,000	216,745	216,970.6	+225.6	216,816

And now we see that even the sun has its spots. Even Gauss's table is "bristling with errors," to use his own phrase. But we should remember that Gauss himself took a more active part only in the primes less than a million and, more importantly, that he himself never published his tables (nor his formula). They were found among his paper after his death.

Gauss apparently never proved his prime number hypothesis. But more than one hundred years later the Belgian de la Vallée-Poussin and the Frenchman Jacques Salomon Hadamard proved, independently of each other, that

$$\pi(x) = \int_2^x \frac{1}{\log n}\, dn - R(x),$$

where the remainder or deviation $R(x)$, which is included in the foregoing table, is relatively small with respect to the first term on the right side of the equality. (This by no means prevents $R(x)$ from taking on arbitrarily large positive or negative values for sufficiently large values of x.) As a matter of fact, we have

$$\pi(x) \sim \int_2^x \frac{1}{\log n}\, dn,$$

where the symbol $\sim$ means that the quotient of the right side divided by the left side approaches 1 as x grows without bound. In general, $f(x) \sim g(x)$ means that the quotient of the two functions $f(x)$ and $g(x)$ goes to 1 as x goes to infinity. One says that $f(x)$ is *asymptotic* to $g(x)$. For example, if we have $f(x) = 2x^3 + x^2$ and $g(x) = 2x^3$, then the quotient is

$$\frac{f(x)}{g(x)} = 1 + \frac{1}{2x}.$$

When x grows without bound, the right side approaches 1, since $1/(2x)$ grows smaller and smaller. Thus we may write $2x^3 + x^2 \sim 2x^3$.

We should observe that in general the relation $f(x) \sim g(x)$ does not imply that the difference $f(x) - g(x)$ goes to 0 as x goes to infinity. In our example we have $f(x) - g(x) = x^2$, which becomes arbitrarily large as x becomes arbitrarily large. The remainder term $R(x)$ in the prime-number formula does the same thing, but it fluctuates in a highly irregular fashion.

Using the relation

$$\pi(x) \sim \int_2^x \frac{1}{\log n}\, dn,$$

one can derive the asymptotic relation

$$\pi(x) \sim \frac{x}{\log x}$$

through relatively simple calculations. Gauss found this formula also, but he correctly thought that it was inferior to the first, since the "error," that is the remainder term $R(x)$, is much larger when one approximates $\pi(x)$ with the latter expression. But this formula is simpler and it is usually called the *prime number theorem*. It implies that the number of primes that are less than or equal to a certain number is about equal to the number itself divided by its natural logarithm. It also implies that if the prime number which has index number n is denoted by P_n, then the relation

$$P_n \sim n \log n$$

holds. During the last fifty years mathematicians have put an enormous amount of work into determining a more accurate estimation of the remainder term $R(x)$. But progress has been minimal in this field to which so much effort has been devoted.

In this connection we ought to point out that the outstanding French mathematician Adrien Legendre (1752–1833) investigated the distribution of prime numbers at about the same time as Gauss. He also used purely empirical methods, and his estimation of $\pi(x)$ was

$$\frac{x}{\log x - 1.08366}.$$

Legendre published this formula in a paper in 1798 and also advanced the hypothesis that

$$\pi(x) \sim \frac{x}{\log x}.$$

This is certainly reminiscent of Gauss's formulas, but each had reached his result independently of the other. Thereafter Gauss and Legendre collided many times in their mathematical research. Neither of them ever explained exactly how they meant the different formulas to approximate $\pi(x)$. It seems likely that they meant an asymptotic relation with a relatively small remainder, something in agreement with the presentation given here.

Finally, as a curiosity, we can relate that in 1963 the American computer Illiac II found that $2^{11213} - 1$ is a prime number. This number has 3,376 digits when written out in the decimal system, and it is the largest prime number known at the present time. Illiac II needed about an hour and a half for its calculations, which included about three fourths of a billion multiplications and additions.

Primes of the form $2^n - 1$ are usually called Mersenne numbers, after the French Franciscan monk Marin Mersenne, who studied them in the early 1600's. The name is inappropriate, since Euclid had already worked with such numbers.

There are two reasons for our having considered Gauss's investigation of the distribution of prime numbers in such detail, even though it constitutes only an insignificant part of his total work. One is that this is Gauss's first contribution as a creative mathematician. It is true that his result was only an unproved hypothesis, but such hypotheses have often played a large role in mathematics, and this one was a fruitful hypothesis which was ultimately found to be correct.

The second reason is that we here see Gauss as a primarily *experimental* mathematician rather than as a theoretician or logician, which is natural considering his youth. The empirical side of Gauss's mathematical work is often neglected for the purely analytical and logical, which is all that he presented in his published work, in which nearly all traces of his labors were wiped away. (This would seem to hold not only for Gauss but to some degree for most mathematicians.)

In his number-theoretical research, and probably in his work in other areas as well, Gauss began with the numbers themselves. He experimented with them from his earliest years, combining them in innumerable ways in numerical calculations—one might even say that he played with them—and during this fine play he empirically found relations and laws whose rigorous proofs later cost him great effort.

Gauss's investigation of the distribution of prime numbers marks the fifteen- or sixteen-year-old's transition to what he himself later called "more subtle research in the higher arithmetic" (which we now call number theory). There his inductive method won its greatest triumph perhaps when he discovered the fundamental theorem of quadratic residues in March of 1795; he called it both *theorema aureum* —the golden theorem—and *gemma arithmeticae*—the gem of arithmetic. But a year passed before he found the first complete proof.

This poetical christening of new theorems perhaps sounds strange to our ears. But Gauss lived during a period that was, from the literary point of view, dominated by romanticism. In his private letters and statements, which were often stamped with formal elegance, he followed the prevalent style. Thus it is romanticism's metaphorical language that is used when he compares mathematical theorems with gold and precious stones during the first decade of the nineteenth century, or when he says "mathematics is the queen of the sciences, but arithmetic is the queen of mathematics." This picture is added to by the fact that, even in his own day, Gauss was called *Mathematicorum princeps.*

To conclude this chapter we shall list several other important ideas and discoveries that Gauss worked on before he went to the university in Göttingen. (We shall come back to them later.)

Already in 1792, when he was fifteen years old, he had begun to ponder the foundations of Euclidean geometry. He was concerned with the famous parallel axiom, and his ideas were later to ripen into non-Euclidean geometry.

In 1791 he started his investigation of the arithmetic-geometric mean. In 1794 he discovered the relation between this average value and certain power series. In the same year he discovered the method of least squares, and he must have studied earlier how to handle observational errors, leading to the Gaussian error curve.

(1736–1813), Frenchman.] Gauss himself also stated that he went to Göttingen rather than to Helmstedt, where the Duchy of Braunschweig's "official" university was located, because Göttingen had a better mathematical library. But the books he borrowed from the university library in Göttingen from October of 1795 to March of 1796 were in most cases not mathematical. Among the twenty-five borrowings there are only five scientific works, among them books of Lambert and Lagrange, while the rest are from the humanities, including for example several volumes of Cicero, Lucian, and Richardson, "Swedish Grammar" by Sahlstedt, and three books of travel written in Swedish: "Travels [to the Cape of Good Hope, etc.]" by Andreas Sparrman, a pupil of Linnaeus, Hasselquist's "Travels to the Holy Land," and Kalm's "Travels to North America I–III." During the years immediately following, however, scientific books dominated those from the humanities.

Thus there are several reasons to support the assertion that Gauss hesitated in his choice of a career. But his matriculation as a student of mathematics does not point toward philology, and probably Gauss had already made his decision when he arrived at Göttingen. He wrote in 1808 that it was noteworthy how number theory arouses a special passion among everyone who has seriously studied it at some time, and, as we have seen, he had found new results in this and other areas of mathematics while he was still at Collegium Carolinum. Further on we shall also quote a couple of statements by Gauss in which he declares his impassioned love for mathematics from the very beginning.

Thus, among the reasons advanced for his hesitation, only the economic one should be taken seriously. But philology remained his hobby. For example, he learned Russian at the age of sixty-four, and learned it well enough so that he could both converse and correspond passably in the language. Gauss borrowed many humanities books during his first year at the university, probably for orientational or recreational reading outside of mathematics, which occupied the largest part of his thoughts.

Posterity has got a glimpse at his mathematical ideas through his scientific journal, which he called *Notizenjournal*, but which was written in Latin. We will quote its first entry—about the 17-gon, which Gauss

The Regular 17-sided Polygon.

The University of Göttingen was founded in 1737 by King George II of England, hence its name Georgia Augusta. It soon took one of the foremost places among German universities, a position it holds to this day.

On October 15, 1795, Gauss was admitted to Georgia Augusta as "matheseos cult."; that is to say, as a mathematics student. But it is often pointed out that at first Gauss was undecided whether he should become a mathematician or a philologist. The reason for this indecision is probably that humanists at that time had a better economic future than scientists. A yearly stipend from the Duke's treasury, paid both in cash and in kind, relieved Gauss of economic troubles for the moment. But one could not know how long that support would continue, and besides Gauss naturally wanted to be independent as soon as possible. In spite of everything he felt constrained by the Duke's benevolence.

The geologist Sartorius von Waltershausen, a close friend of Gauss in later years, said in his memorial paper on Gauss that Gauss first became completely certain of his choice of studies when he discovered the construction of the regular 17-sided polygon; that is to say, after his first year at the university. Of the professors at Göttingen, Gauss was more impressed by the great philologist and classicist Christian Gottlob Heyne, whose lectures he attended at first, than he was by the mathematician Abraham Gotthelf Kästner, who showed but little understanding for Gauss's research.

Gauss's library loans during his first year at Göttingen are surprising. As a student at Collegium Carolinum he had probably studied Newton, Euler, and Lagrange thoroughly, and it would have been natural for him to continue along that line. [Isaac Newton (1642–17?), Englishman. Leonard Euler (1707–1783), Swiss. Joseph-Louis Lagra considered among his most important discoveries. The journal begins on March 30, 1796, when he was eighteen years old, with the notation:

> *Principia quibus innititur sectio circuli, ac divisibilitas eiusdem geo-metrica in septemdecim partes etc.*
>
> Mart. 30. Brunsv[igae]

> [Principle of the circle's division, and how one geometrically divides the circle into seventeen parts, and so forth.
>
> March 30, Braunschweig]

It was this discovery that made Gauss famous among mathematicians in a single stroke. It involves construction of the regular 17-sided polygon with only "Euclidean tools," that is to say, with compass and straightedge. Gauss reported how he found the solution:

> Through deep reflection upon the arithmetical relation between the roots [of the equation $(x^p - 1)/(x - 1) = 0$] during a vacation in Braunschweig, on the morning of the day in question (before I had gotten out of bed), I managed to see the desired relationship with such great clarity that I could immediately carry through the special application to the 17-gon and confirm it with a numerical calculation.

Gauss was so happy about and so proud of his discovery that he told his friend Wolfgang Bolyai that the regular 17-sided polygon should be carved into his gravestone. It was not to be, but on the monument to Gauss in Braunschweig there is a seventeen-pointed star; the sculptor who carried out the work maintained that everyone would confuse a regular seventeen-sided polygon with a circle. In this idea, Gauss followed a classical tradition. According to tradition, Archimedes wanted to have a sphere inscribed in a cylinder on his gravestone (Cicero found such a stone, overgrown with vines, when he was quaestor of Sicily), and Jacob Bernoulli has a logarithmic spiral on his gravestone. Gauss announced his discovery in the "Intelligenzblatt der allgemeinen Litteraturzeitung" on June 1, 1796. The notice, which is Gauss's first appearance in public print, gives a historical survey of the problem.

> Every beginner in geometry knows that it is possible to construct different regular polygons, for example triangles, pentagons, 15-gons, and those regular polygons that result from doubling the number of sides of these figures. One had already come this far in Euclid's time, and it seems that since then one has generally believed that the field for elementary geometry ended at that point, and in any case I do not know of any successful attempt to extend the boundaries beyond that line.
>
> Therefore it seems to me that this discovery possesses special interest, *that besides these regular polygons, a number of others are geometrically constructible, for example the 17-gon.* This discovery is really only a corollary of a theory with greater content, which is not complete yet, but which will be published as soon as it is complete.
>
> *C. F. Gauss, Braunschweig*
> Mathematics Student at Göttingen.

This is Gauss's first and only contribution to any journal of advance notices (very common today), which are used for securing priority rights to a scientific discovery.

The complete theory behind the construction of the regular 17-gon lies beyond the scope of this book. We shall point out only how it begins. But first let us say a bit about complex numbers and how one interprets them geometrically. This is a part of mathematics that also has close connections to Gauss.

Every extension of the notion of number has met great opposition because of our intuitive inclination to see a direct connection between numbers and concepts based upon practical geometry, above all the concept of the distance between two points. Negative numbers had a much harder time breaking through than did positive irrational numbers, such as $\sqrt{2}$, which shows up when one calculates the length of the diagonal of a square of side 1 using the Pythagorean theorem. Instead, the negative numbers were introduced via algebra, where they were needed from the standpoint of completeness, so that, for example, one could say that a polynomial equation of the first degree always has a root. But long into the seventeenth century negative numbers had their opponents, who did not wish to recognize this extension of the notions of number; and rather than say that the equation $x + 5 = 0$ has a negative root, namely -5, they preferred to say that it had no root at all.

The "imaginary" numbers met with similar difficulties, even the name indicating that they have to do with something that has no existence in the real world. After one accepts the fact that the equation $x^2 - 1 = 0$ has two roots, namely $+1$ and -1, one encounters a new problem in the equation

$$x^2 + 1 = 0,$$

which has no roots in the domain of real numbers. In order for it to have any roots at all, we are again compelled to extend the domain of numbers. If we purely formally write $x^2 = -1$ and then take square roots of both sides, we have the roots

$$x_1 = +\sqrt{-1} \text{ and } x_2 = -\sqrt{-1}.$$

Euler (and others before him) denoted $\sqrt{-1}$ with the symbol i, an abbreviation for the word "imaginary." The usual rules of arithmetic hold for i, including the rule $i \times i = -1$, or $i^2 = -1$. With the use of this symbol one finds that $x^2 + 1 = 0$ has the roots $x_1 = i$ and $x_2 = -i$. In like manner one finds that the roots of the equation $x^2 - 2x + 5 = 0$ are $x_1 = 1 + 2i$ and $x_2 = 1 - 2i$. In general we can now say that every polynomial equation of the second degree has two roots.

A number of the form $a + ib$, where a and b are arbitrary real numbers, is called, according to Gauss's suggestion, a *complex number*, while in the special case in which $a = 0$, the number ib is called a *pure imaginary number*. The numbers $1 + 2i$ and $\sqrt{3} + i\sqrt{2}$ are complex numbers, while i and $i\sqrt{5}$ are pure imaginary numbers.

Thus the complex numbers were introduced in a purely algebraic way. The geometric illustration was harder in this case than for the negative numbers, for which no doubt the interpretation became clear to all after negative degrees were introduced on the thermometer.

Help came from a geometric problem, which had already been solved by Euclid: "Construct the proportional (or geometric mean) of two given distances a and b." The solution is to take a line segment of length $(a+b)/2$, which is the arithmetical average or just plain average of a and b, as the radius of a semicircle (Fig. 3).

From point D we draw a normal to AB, which intersects the circle at C. The triangle ACD is a right triangle, since every angle inscribed in a semicircle is a right angle. Since the triangles ACD and BCD are similar, we have

$$\frac{a}{h} = \frac{h}{b},$$

and thus

$$h^2 = ab \text{ and } h = \pm\sqrt{ab}.$$

Since h is a distance, we choose the positive sign. Then $CD = h$ is the sought for *proportional* $\sqrt{ab}$. From the triangle OCD we then see that r is equal to or greater than h (written $r \geq h$); that is to say,

$$\frac{a+b}{2} \geq \sqrt{ab},$$

and the equality sign holds only if $a = b$. We can use this special case to illustrate the imaginary unit i. On the real-number line we choose

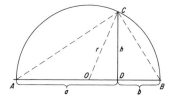

Figure 3
Euclid's construction of the proportional or geometric mean value of $h = \sqrt{ab}$ of two given distances a and b.

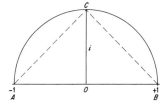

Figure 4
Wessel's method of illustrating the imaginary unit $i = \sqrt{-1}$.

the points $x = +1$ and $x = -1$, and draw a semicircle as in Fig. 4.

In analogy with the foregoing we now have

$$OC = \sqrt{(-1) \times (+1)} = \sqrt{-1} = i.$$

We should remember that this is not a purely geometric interpretation, since a line or a distance cannot be negative. But formally we can say that i is the proportional to two coordinates on the x axis, namely $+1$ and -1.

The Norwegian-Danish surveyor Caspar Wessel published this method of illustrating imaginary numbers in a paper in 1799; however, it did not attract any great attention. The Frenchman Jean-Robert Argand developed the same notion in 1806. Gauss probably did not know of these works, in any case not Wessel's, but he seems to have arrived at the same idea about 1800. He later gave up the geometric picture and gave the first purely logical presentation of the complex numbers as number pairs, for which the elementary rules of calculation are defined in a specific way.

Gauss developed the idea of Fig. 4 into the *complex plane*, which is also called the *Gaussian number plane*. The y axis in the usual coordinate system is replaced with the imaginary axis (Fig. 5). Just as in the usual coordinate system, a point is here defined by two real numbers. But we now write $z = x + iy$, where x and y are real numbers and i is the imaginary unit. For example, we write $z = 3 + 2i$ for point A, $z = 2 - 3i$ for point B, and generally $z = a + ib$ or $z = x + iy$ for an arbitrary point p. Real numbers $z = a$ lie on the x axis, and pure imaginary numbers $z = ib$ on the y axis. To every point in the Gaussian plane there corresponds a complex number, and conversely to every complex number there corresponds a point in the Gaussian plane.

Now one usually interprets i geometrically as a *rotation*. If we multiply the number 1 by i we obtain the number i; that is to say, we get i by rotating the segment between the origin 0 and $+1$ through an angle of 90° in the positive direction, that is, the counterclockwise direction. If we then multiply i by itself we get $i \times i = -1$; in other words, we obtain -1 by rotating the segment between 0 and $+i$ through an angle of 90°, or else by rotating the segment between 0 and 1 through 180°. In the same way we then find that $i \times i \times i = i \times i^2 = -i$ and $i^4 = 1$. For an arbitrary point $P = a + ib$ we have (Fig. 6)

$$(a + ib) \times i = ai + i^2 b = -b + ia.$$

Using elementary geometry, we see that the angle POQ is a right angle, and that we go from point $P = a + ib$ to point $Q = i \times P = -b + ia$ by rotating the segment OP through an angle of 90°.

Since the complex numbers have been introduced, the construction of a regular n-sided polygon can be discussed with help of the binomial (two-termed) equation $z^n = 1$, where n is a natural number. The solutions of these equations for $n = 3$ and $n = 4$ are obtained by using the so-called cube and conjugate rules. We then find for $n = 3$ that

$$z^3 - 1 = 0 \text{ and } (z - 1)(z^2 + z + 1) = 0,$$

$-\frac{1}{2} + \frac{\sqrt{3}}{2}i \qquad -\frac{1}{2} - \frac{\sqrt{3}}{2}i$

Error

which gives us the roots $z_1 = 1$, $z_2 = \left(-\frac{1}{2} + i\sqrt{\frac{3}{2}}\right)$ and $z_3 = \left(-\frac{1}{2} - i\sqrt{\frac{3}{2}}\right)$. For $n = 4$ we have $z^4 = 1$, so that $(z^2 - 1)(z^2 + 1) = 0$, which gives us the roots $z_1 = 1$, $z_2 = i$, $z_3 = -1$, and $z_4 = -i$. If we plot the corresponding points in the complex plane we have the vertices of an equilateral triangle and a square, respectively. Both figures can be in-

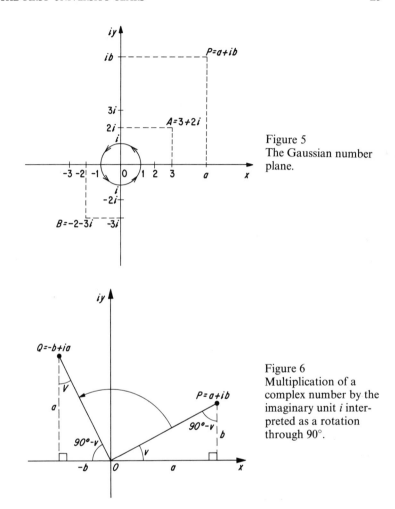

Figure 5
The Gaussian number plane.

Figure 6
Multiplication of a complex number by the imaginary unit i interpreted as a rotation through 90°.

scribed in a circle around the origin with radius 1, the so-called *unit circle* (Figs. 7 and 8).

One can also express the roots in trigonometric form. If an arbitrary point $P = a+ib$ lies on the unit circle, we have

$$P = \cos v + i \sin v,$$

since $a = \cos v$ and $b = \sin v$ (Fig. 8). We see that the roots of the equation $z^3 = 1$ correspond to $v = 0°$, $v = 120°$ and $v = 240°$ (Fig. 7), and that the roots of the equation $z^4 = 1$ correspond to $v = 0°$, $v = 90°$, $v = 180°$ and $v = 270°$ (Fig. 8).

The roots of the two equations can be written in trigonometric form as follows:

$z_1 = \cos 0° + i \sin 0°$
$z_2 = \cos 120° + i \sin 120°$
$z_3 = \cos 240° + i \sin 240°$

$z_1 = \cos 0° + i \sin 0°$
$z_2 = \cos 90° + i \sin 90°$
$z_3 = \cos 180° + i \sin 180°$
$z_4 = \cos 270° + i \sin 270°$

The first sequence, which gives the roots of $z^3 = 1$, is usually given by the shorter notation

$$z_k = \cos k \times \frac{360°}{3} + i \sin k \times \frac{360°}{3}, \, k = 0, 1, 2.$$

By this notion we mean that the index k takes on the values 0, 1, and 2.

In the same way the second sequence which gives the roots of $z^4 = 1$ can be written as

$$z_k = \cos k \times \frac{360}{4} + i \sin k \times \frac{360}{4}, \, k = 0, 1, 2, 3.$$

From the examples given here we can conjecture that the roots of the equation $z^n = 1$, the so-called nth roots of unity, are

$$z_k = \cos k \times \frac{360°}{n} + i \sin k \times \frac{360°}{n}, \, k = 0, 1, 2, \ldots, n-1.$$

Using a theorem which has been named for the Frenchman Abraham de Moivre, one can prove that this conjecture is correct.

As mentioned before, construction of a regular n-sided polygon in the sense of Euclidean geometry restricts one to the use of compass and straightedge. It can be proved that it is possible, under these restrictions, to construct only numbers that are built from rational numbers, square roots, or a finite number of rational number operations and square root computations. Thus cube roots, in general, cannot be constructed with straightedge and compass; it is for this reason that the ancient

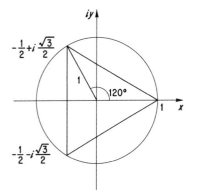

Figure 7
Roots of the equation $z^3 = 1$ illustrated in the Gaussian number plane.

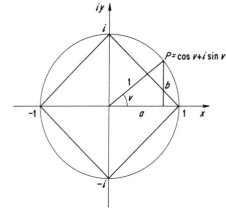

Figure 8
Roots of the equation $z^4 = 1$ illustrated in the Gaussian number plane.

problems of trisecting the angle and doubling the cube are unsolvable except in special cases.

In the problem of dividing the circle we know, thanks to the binomial equation $z^n - 1 = 0$, that we can construct a regular n-sided polygon if we can express $\cos 360°/n$ with the help of a finite number of rational numbers or square roots. In principle we can start with a unit circle and construct the line $\cos 360°/n = OP$ with straightedge and compass (Fig. 9). From P we then draw a normal that intersects the circle at B. We find that $BP = \sin 360°/n$ and that the point B in the complex plane is determined by

$$B = \cos\frac{360°}{n} + i\sin\frac{360°}{n};$$

that is to say, it corresponds to the root of unity we obtain when we set $k = 1$. Of course, point A corresponds to $k = 0$, and thus AB is one side of the regular n-sided polygon, which can then be easily completed with the compass. From the algebraic viewpoint such a construction corresponds to reducing the solution of the binomial equation $z^n - 1 = 0$ to the solution of a series of equations of first or second degree.

The ancient Greeks were able to solve the cases in which $n = 3$, $n = 5$, $n = 2^m$, where m is a natural number, and the cases that follow from these, namely, 15, 3×2^m, 5×2^m and 15×2^m. Thus, in addition to the examples named earlier—the triangle and the square—the Greeks were able to construct the regular pentagon, which has played a large role as a magic symbol, from the Pythagoreans to the U.S.A.'s Pentagon Building and the Soviet Union's Star.

In the case of the regular pentagon,

$$\cos\frac{360°}{5} = \cos 72° = \frac{1}{4}(\sqrt{5} - 1),$$

and $\sqrt{5}$ can be constructed as the proportional to the distances 1 and 5. From this we easily obtain a line of length $\cos 72°$. The regular pentagon can then be constructed according to Fig. 9, if we choose $n = 5$ there (see also Fig. 10).

From this point on no progress was made for about 2,000 years, until the eighteen-year-old Gauss took the problem under consideration. He knew the solution of the binomial equation $z^n - 1 = 0$ in trigonometric form. The general theory, which he mentioned in his journal notice cited earlier, had as an especially interesting case the value $n = 17$.

The roots of $z^{17} - 1 = 0$ are

$$z_k = \cos k \times \frac{360°}{17} + i\sin k \times \frac{360°}{17}, k = 0, 1, 2, \ldots, 16,$$

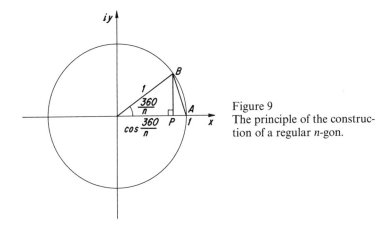

Figure 9
The principle of the construction of a regular n-gon.

Figure 10
The pentagram, constructed from diagonals of a regular pentagon, has always been a magic symbol.

where $k = 0$ and $k = 1$ correspond to A and B in Fig. 9. Gauss eliminated the trivial root $z_0 = 1$ by dividing by the factor $z-1$. He then had

$$\frac{z^{17}-1}{z-1} = 0.$$

After the division is carried out the equation becomes

$$z^{16}+z^{15}+ \ldots +z+1 = 0,$$

which has 16 roots that are obtained by setting $k = 1, 2, \ldots, 16$. The deciding factor in constructibility now is whether or not the solution of this equation can be reduced to the solution of a sequence of equations of the first or second degree. If so, then the trigonometric expression

$$\cos \frac{360°}{17}$$

can be written in the form of an expression that contains only a finite number of rational numbers and square roots. Gauss showed that this was possible, and if we compare his result with the simple expression we presented above for the regular pentagon, the difference in degree of difficulty appears overwhelming. Gauss found that

$$\cos \frac{360°}{17} = -\frac{1}{16} + \frac{1}{16}\sqrt{17} + \frac{1}{16}\sqrt{34 - 2\sqrt{17}}$$

$$+ \frac{1}{8}\sqrt{17 + 3\sqrt{17} - \sqrt{34 - 2\sqrt{17}} - 2\sqrt{34 + 2\sqrt{17}}}.$$

This is unquestionably an expression that contains only a finite number of rational numbers and square roots, and thus it can be constructed with straightedge and compass. Gauss also calculated the numerical values of the roots of unity $z_1, z_2, \ldots, z_{16}$. They can be paired together two by two. The first pair is

$$z_{1,16} = \cos \frac{360°}{17} \pm i \sin \frac{360°}{17},$$

or

$$z_{1,16} = 0.9324722294 \pm 0.3612416662i$$

approximately, to ten decimal places. Gauss also calculated the other roots to ten decimal places.

This in itself is an astounding performance, but Gauss did even more. For a general investigation of constructibility it is clearly sufficient to consider only odd numbers, since an arc of a circle can easily be divided in two with a compass—for example, the regular 10-sided polygon is obtained by dividing the arcs outside a pentagon. Gauss extended the ancient series mentioned above and proved the following:

The construction with straightedge and compass of a regular polygon with an odd number p of sides is possible if p is either a prime number

of the form $2^{(2^m)}+1$ or a product of different primes of this type. Here
m is any of the numbers $0, 1, 2, \ldots$.

Prime numbers $2^{(2^m)}+1$ are called *Fermat primes* after the great
French mathematician Pierre Fermat (1601–1665), who investigated
them and hypothesized that all numbers of the form $2^{(2^m)}+1$ are
prime numbers. This is not the case. If we set $m = 0, 1, 2, 3, 4, 5$ we
obtain the numbers

   3, 5, 17, 257, 65537, 4294967297.

The first five numbers are primes, but the sixth admits, as Euler later
showed, the factor 641. The subsequent investigation of these numbers
for different values of m have resulted in the discovery of no Fermat
prime larger than 65,537. As a curiosity we mention that in 1958 it
was found by use of a computer that $2^{2^{1945}}+1$ is divisible by
$5 \times 2^{1947}+1$. The first of these numbers is incomprehensibly large; if
one wrote it out in the decimal system using numerals one centimeter
in width, it would reach around the now observable universe an un-
imaginably large number of times.

   The regular 257-sided polygon has been investigated by Richelot,
who carried out the construction in a paper of 194 pages. Professor
Hermes devoted ten years of his life to a detailed investigation of the
regular 65,537-gon. The results of his learned labors are preserved in a
trunk in the attic of the Mathematical Institute at Göttingen.

   We see that the regular polygons with 7, 9, 11, and 13 sides cannot
be constructed with compass and straightedge. On the other hand, one
can construct regular polygons with $3 \times 17 = 51$, $3 \times 5 \times 17 = 255$,
$2 \times 5 \times 65,537 = 655,370$ sides, and so forth. The general result can
thus be formulated:

The only regular n-sided polygons that can be constructed with compass and straight-edge are those for which

$$n = 2^k \times p_1 \times p_2 \times \ldots, k = 0, 1, 2, \ldots,$$

and $p_1, p_2, \ldots$ are distinct Fermat primes.

To conclude this section we give a construction of the regular 17-gon with compass and straightedge, as given in "Gelöste und ungelöste Probleme aus alter und neuer Zeit" by Heinrich Tietze (Munich, 1949). In order to describe the method we use analytic geometry. The construction is not as difficult as it looks, but the proof that it is correct requires extensive calculations.

The equation for a circle with center at the point with coordinates (a,b) and with radius r is, according to the Pythagorean theorem (Fig. 11)

$$(x-a)^2 + (y-b)^2 = r^2,$$

where (x,y) are coordinates of an arbitrary point on the circle.

In the construction the circles that appear will be written in this form. If $r = \sqrt{1+d^2}$, then we may infer that the radius of the circle is the hypotenuse of a right triangle whose two smaller sides have lengths 1 and d.

Let the given circle be the unit circle $x^2 + y^2 = 1$ (Fig. 12). It cuts the positive y axis at $A = (0,1)$ and the negative x axis at $B = (-1,0)$. Draw the circle $x^2 + (y - \frac{1}{4})^2 = \frac{17}{16}$. It has its center at $C = (0,\frac{1}{4})$, cuts the negative x axis at B, the positive y axis at $D = (0,d)$, and the negative y axis at $E = (0,e)$. The circle $x^2 + (y-d)^2 = 1+d^2$ goes through B and cuts the negative y axis at $F = (0,f)$. The circle $x^2 + (y-e)^2 = 1+e^2$ goes through B and cuts the negative y axis at $G = (0,g)$. Find the midpoint $H = (0,h)$ of the line segment AF. The circle $x^2 + (y-h)^2 = (1-h)^2$ cuts the positive x axis at $I = (i,0)$. Let $K = (2i,0)$. The circle $(x-2i)^2 + y^2 = g^2$ cuts the positive y axis at $L = (0,l)$. The line $y = \frac{1}{4}(l-g)$ cuts the given circle at M and N. The line is parallel to the x axis at a distance $\frac{1}{4}(l-g)$. The points M, A, and N are three of the successive vertices of a regular 17-gon inscribed in the unit circle.

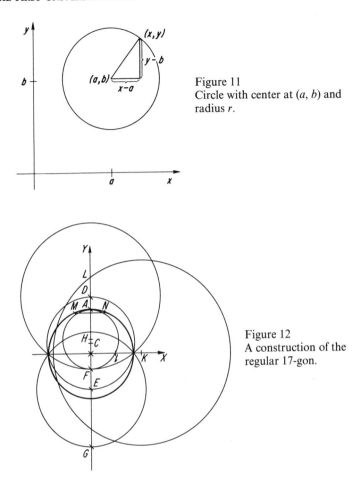

Figure 11
Circle with center at (a, b) and radius r.

Figure 12
A construction of the regular 17-gon.

The Journal. Gauss's Claim to a Mathematical Result.

Gauss's journal was not found until 1898. It plays an important role in judging his mathematical contributions, since it contains a great deal that Gauss never published or which he only hinted at in letters to his friends. It is a booklet of nineteen small octavo pages, covering the period from March 30, 1796 to July 9, 1814. Altogether there exist 146 very short accounts of discoveries, results of numerical calculations, or simple assertions of mathematical theorems. For example, the notice of March 30, 1796, given earlier, is all we find about the construction of the regular 17-gon.

The journal gives us a clear view of Gauss's mathematical development, particularly during the significant years 1796–1801, when Gauss was 18 to 24 years old—almost all of the entries, 121 of them, fall in this interval. We can, in a laconic form, follow a string of great discoveries in algebra, analysis, and number theory. The discoveries in number theory led to the publication of his "Arithmetical Investigations."

In the journal, Gauss has laid aside his mask of caution and inaccessibility which he showed to the world around him during his entire life. We see here his own reactions to his discoveries. His joy and pride break forth in triumphant exclamations. "*Felicitas nobis est facta*"—I have succeeded—closes the notice of June 3, 1800, when he found a most beautiful property of the elliptic modular function. Similar expressions appear several times and are reminiscent of Archimedes' "Eureka"—I have found it—the triumphant cry of the discoverer throughout the ages.

Between the great discoveries one sometimes finds simpler results inserted, showing the basis of numerical computations or trivial calculations. In a couple of places there are notations that consist of short cryptograms. Which giant had Gauss defeated on October 21, 1796, when he wrote "*Vicimus Gegan*"—we have defeated Gegan? There is one further such entry, which no one has been able to decipher, but the other entries have been the subject of a careful exegesis in his collected work.

In spite of the laconic methods of expression in the journal, Gauss's individuality shows forth: the rich ingenuity in combination with numerical proficiency, and the logical power to find exact proofs after long and laborious work. It was this combination that led Gauss to triumph where others had failed.

As with Mozart, Gauss is supposed to have been so overwhelmed with new ideas during his youth that he did not have time to carry out one task before another turned up. The journal entries were made in Latin. Probably it was Gauss's intention to put them into detailed form later. But in many significant cases Gauss never published his discoveries, which instead were recovered from the notes he left behind or from his letters. For posterity, and certainly for our time, his behavior seems hard to understand. We shall try to give an explanation.

Where a scientific discovery is concerned the earliest publication date determines the right of priority, and it can also determine immortal renown. In our day the conflict in this area is very severe. Can one imagine Gauss supplying an average of two notices per month for fifty years and thereby securing priority rights to the most important mathematical discoveries of that period?

One cannot imagine it. He certainly was not a complete stranger to such methods. He announced the construction of the 17-gon five years before he published the proof, and he was certainly no less impressed with fame than other men. But one still cannot imagine it, since Gauss measured the worth of published works by other standards than those of most men. In this and in many other cases he reminds us of Newton, whom he admired most of all. While other mathematicians in Gauss's writings are labeled with epithets such as *"clarissimus"*—highly distinguished, Newton is *"summus"*—the very best.

Gauss was also impressed by the logical rigor of the Greek mathematicians—*"rigor antiquus."* It had influenced Newton, who in his *Principia* often used *infinitesimal calculus* to derive his theorems, and then wiped away all traces of the calculus and used a method of proof with Euclid as his guide in the published version.

It was only toward the end of the 18th century that mathematicians began to give serious attention to logical rigor in the theory of functions. Gauss was completely devoted to that line, but since he published little of his work in analysis, the role of pioneer among the "rigorists" is awarded to such great mathematicians as the Frenchman Augustin Cauchy and the Norwegian Niels Abel.

Gauss's incorruptible insistence on rigor made the path from an intuitive form of a theorem to the complete mathematical proof very difficult. Because of this Gauss made the severest demands on himself, even from a purely formal point of view.

Like Archimedes and Newton, he wished to leave behind only complete works of art, in which nothing could be changed without destroying the inner harmony. He used to say that when a building is completed no one should be able to see any trace of the scaffolding. Many mathematicians since have certainly wished they could see the traces; it would have made his work easier to read.

To the demands for rigor and formal clarity, Gauss added another which he had in common with Newton: the demand for synthesis. He wished to give a uniform and general theory in each area, one that gave the connections between the different theorems.

Keeping in mind that Gauss worked in all the areas of mathematics and its applications of his time, these three demands together comprise a superhuman program. To what extent he would have been able to realize such a program we do not know. The first condition for such a task is that he should have had what we now call a purely research professorship in mathematics. But such a thing did not exist at that time.

Again, like Newton, Gauss wanted to obtain a secure social position. Duke Ferdinand certainly supported him without any conditions, but Gauss wanted to stand on his own feet. The post as chief of the astronomical observatory in Göttingen led him to economic independence in 1807, but also to tens of thousands of astronomical and geodetic measurements, which could just as well have been carried out by someone lacking his creative genius.

A further reason for his reserve is to be found in Gauss's character. He detested violence of all forms and looked upon impetuous outbursts as disgraceful. Probably this attitude was a reaction against his tyrannical father. If Gauss had published his ideas about imaginary numbers or non-Euclidean geometry, for example, a public debate with the attendant criticism would have been unavoidable, especially in the question of the new geometry and its defiance of "common sense."

Gauss held a profound dislike for polemics, which so easily degenerates into what he called "Boeotians' cries" in a famous statement—that is to say, the asinine human's cry. He also wanted to protect his peace of mind for his work. He did not take part in public debate, although he could be sharp in his criticism in private discussions with his friends.

Thus there are several reasons that so many of the brilliant discoveries in the journal were never published by Gauss himself. The most important are certainly his demands for rigor, beauty, and synthesis. It is these demands that found expression in Gauss's seal. It contains among other things a tree bearing only seven fruits. The statement beneath reads: *Pauca sed Matura*—"Few but Ripe."

When one sees—or lifts—Gauss's collected works, in twelve large and weighty volumes, one cannot fail to consider the seal indicative of Gauss's accomplishment in pure mathematics.

Friendship with Wolfgang Bolyai. The Fundamental Theorem of Algebra. Gauss as a Rigorist.

From his earliest childhood, probably because of his role as a prodigy and also perhaps because of his serious disposition, Gauss became isolated from his contemporaries. This solitude continued during his first years at Göttingen. Gauss did not take part in student life and he had but few friends.

Among his friends was Wolfgang Bolyai von Bolya (1775–1856), whom we mentioned earlier. Bolya was a Hungarian nobleman, born on the Bolya estate, about 250 kilometers north of Hermannstadt in Transylvania (where one of the great pioneers of astronautics, Hermann Oberth, was born in 1894). Bolyai studied at Göttingen during 1796–99 and later became professor of mathematics at Maros-Vásárhely in Transylvania.

In a letter Bolya reports: "The passion for mathematics (not outwardly shown) and our spiritual compatibility bound us together so that often, when we were out walking, we could stroll quietly for hours, each of us taken up with his own thoughts."

But one should not draw the conclusion from this that there was no exchange of ideas between the two students. Gauss said later that Bolyai was the only one who understood his view of the foundations of mathematics. Bolyai's statement is instead a proof of their friendship, since surely one of the tests of friendship is the ability to be silent together for a long period.

When they talked with each other, they often discussed Euclid's axioms, and Wolfgang both surprised and pleased Gauss with his definition of a straight line and with his different attempts to prove the parallel axiom. These were among the problems that Gauss himself was mulling over at the same time.

In the year 1797 Gauss and Bolyai journeyed on foot to Braunschweig to call on Gauss's parents. Mama Dorothea understood her son's studies and their aims even less than Papa Gebhard. But her heart was with her son. Once when Gauss was out she asked Bolyai what was to become of her son. Upon hearing the answer, "He is Europe's greatest mathematician," she burst into tears.

Gauss and Bolyai exchanged letters for more than fifty years, from 1797 to 1853. These letters give us a vivid picture of both the highly imaginative Hungarian and of the warm and faithful friendship which Gauss, in spite of his icy temperament, was able to harbor. The most noteworthy fact about Wolfgang Bolyai, however, was not that he was a close friend of Gauss but that he was the father of Johann Bolyai, one of the founders of non-Euclidean geometry. We shall return to Johann later.

During his first student years Gauss appeared not only in Göttingen and Braunschweig but also in Helmstedt, where Johann Friedrich Pfaff (1765–1825) was professor of mathematics. In this university city there was a fine mathematical library, but it was certainly more important that Pfaff was there, since Gauss got along much better with him than he did with Kästner. At that time Pfaff was Germany's best-known mathematician, and Gauss found him worthy of his reputation.

Gauss published his first scientific work in Helmstedt in 1799, his doctoral thesis *Demonstratio nova theorematis omnem functionem algebraicum rationalem integram unius variabilis in factores reales primi vel secundi gradus resolvi posse*—a new proof that every polynomial of one variable can be factored into real factors of the first or second degree.

On July 16, 1799, in his absence, he was awarded his Doctor of Philosophy degree at the university in Helmstedt. His thesis is a proof of the fundamental theorem of algebra, which is related to the factor theorem of high-school algebra. We shall illustrate it with several examples of integral rational functions $f(x)$, that is to say polynomials, and their factorization. The problem of breaking up a polynomial $f(x)$ into factors is practically the same as finding the roots of the equation

$$f(x) = 0.$$

Let us first consider

$$f(x) = x^2 - 3x + 2,$$

and solve the second-degree equation

$$x^2 - 3x + 2 = 0,$$

which has the roots $x_1 = 1$ and $x_2 = 2$. We can now break up the polynomial into two real factors and find

$$(x^2 - 3x + 2) = (x - 1)(x - 2).$$

If we next look at

$$f(x) = x^2 - 10x + 25,$$

then we must solve the equation

$$x^2 - 10x + 25 = 0,$$

and we get the roots $x_1 = 5$ and $x_2 = 5$; that is to say, we get the same root twice. The terminology in this case was unclear in Gauss's time, but now we say that the equation has a *double root* or *two equal roots*, so that this second-degree equation also has two roots (and not just one). Factoring it yields

$$x^2 - 10x + 25 = (x-5)(x-5) = (x-5)^2.$$

As a final example we consider the equation

$$f(x) = x^4 - 1.$$

From the chapter on the 17-gon we recall that the roots of the equation are ± 1 and $\pm i$, and the factorization is

$$f(x) = (x+1)(x-1)(x^2+1).$$

We have two real factors of the first degree and one real factor of the second degree. As is apparent from its title, this is the way Gauss carried out the factorizations in his thesis. He consistently avoided imaginary numbers since at that time they were a seed of dissension among mathematicians. In his careful way he worked in the usual x–y coordinate plane at all times, without indicating anywhere that in fact he was giving the geometric interpretation of the complex numbers $x + iy$.

According to his journal Gauss found the proof in October of 1797. He later found three more proofs, the last in 1849 at the time of the celebration of the golden jubilee of his doctorate. In this fourth proof, which is really only a worked-over version of the first, he at last openly operated with complex numbers, since, as he then stated, they "are nowadays apparent to all."

Thanks to the complex numbers it is possible to give factorization in factors of only the first degree, and we can write for example:

$$x^4 - 1 = (x+1)(x-1)(x+i)(x-i).$$

Another advantage is that we may now allow the coefficients of the polynomial to be complex numbers.

The fundamental theorem of algebra may now be stated in its general form: *Every polynomial equation has at least one root.* The fact that a polynomial equation of degree *n* always has *n* roots is then a simple corollary. Without complex numbers the unity and lucidity are lost, and the theorem falls into various special cases.

Usually we look upon it as a relatively obvious fact that a second-degree equation should have two roots, a third-degree equation, three roots, and so on for equations of arbitrary degree. But how do we *know* that a general equation of degree 127, for example, really has 127 roots, or even one root? The assertion requires an *existence proof*, that is to say, one must give a proof that every algebraic equation really has a root. This gave rise to the name "fundamental theorem of algebra," and in his doctoral dissertation Gauss gave one of the first and most important contributions to a long series of existence proofs in different branches of mathematics.

Gauss called his thesis "A *new* proof . . . ." He certainly did not do this because of any shyness or respect for his predecessors, to whose results he refers. In the formulation there is also some indication that Gauss knew that his proof itself was not logically unassailable. It was certainly more rigorous than its predecessors, but Gauss accepted as self-evident certain properties of what are called continuous functions, which in fact were proved later for the first time by the Austrian Bernard Bolzano (1781–1848). This explains both the title and the fact that he repeatedly returned to the same problem.

The different proofs of the fundamental theorem of algebra are Gauss's most important contributions as a rigorist, that is to say, as a representative of logical strictness in method of proof. Since this theorem has great significance in both algebra and function theory, it has influenced both areas.

But the stimulation to rigor came most of all from function theory. During the seventeenth and eighteenth centuries mathematicians, with the fresh joy of discovery, had sprung out upon the new expanses opened up by the differential and integral calculus. They had dealt with infinite series and with "infinitesimals" in a purely formal manner, without seriously asking themselves if these manipulations with infinitely many quantities or with "infinitely small quantities" really had any meaning. In the hands of geniuses such as Newton, Leibniz, Euler, and members of the Bernoulli family, everything had gone amazingly well, and the results were brilliant. [Gottfried Wilhelm Leibniz (1646–1716) was German. The Bernoulli family were Swiss. The best known members are Jacob (1654–1705), Jean (1667–1748), and Daniel (1700–1782).]

Toward the end of the eighteenth century there were second thoughts in the form of increasing criticism of the absurdities that result when, for example, one does not inquire whether an infinite series has any logical meaning. Let us illustrate this problem with a simple case. If we divide the number 1 by the expression $(1 - x)$ we obtain a purely formal series development that begins as follows:

$$\frac{1}{1-x} = 1 + x + x^2 + x^3 + \ldots .$$

The question now becomes: for which values of x does this expression have any meaning, if we suppose that the series on the right side continues without end?

If in turn we choose $x = 2$, $x = -1$, and $x = 1/10$ we obtain

$$-1 = 1 + 2 + 4 + 8 + \ldots ,$$

$$\frac{1}{2} = 1 - 1 + 1 - 1 + \ldots , \text{ and}$$

$$\frac{10}{9} = 1 + \frac{1}{10} + \frac{1}{100} + \frac{1}{1000} + \ldots .$$

Of these three expressions the first is obviously unreasonable, since the right side consists of only positive numbers and grows without bound as the number of terms increases. If in the second expression we compute the sum of an even number of terms on the right side it will always be equal to 0, while the sum of an odd number of terms will always be equal to 1. This is not altogether satisfying, even though Leibniz thought that for probability's sake this infinite sum should be just $\frac{1}{2}$, since the sum of an arbitrary number of terms has equal likelihood of being 0 and 1, and the number $\frac{1}{2}$ lies halfway between them!

However, this reasoning is not much better than metaphysical postulates such as the assertion that, if infinity is an even number, the right side is 0, while if infinity is an odd number, the right side is 1. Because, after all, infinity is not a number. The only thing we can say about infinity is that no matter how large a number we choose, infinity is *larger*. Thus it cannot be a number in the usual sense. It is usually denoted by the symbol ∞.

Unlike the others, the third expression agrees with our intuitive notion of an infinite series having meaning, and by computing the sum of more and more terms on the right side we find directly that it comes nearer and nearer to 10/9. Thus 10/9 can be expressed by the infinite decimal expression 1.11111 . . . .

The reader who is familiar with infinite series knows the answer to the question posed: the infinite series $1 + x + x^2 + \ldots$ defines a number, provided x takes on a value between -1 and 1; this condition is written in abbreviated form as $-1 < x < 1$. In this case the series is *convergent*, that is to say, the sum on the right approaches a definite finite value when the number of terms grows without bound.

As an eleven-year-old, Gauss was already studying Newton's binomial theorem, which includes the infinite geometric series as a special case. He investigated the conditions under which an infinite binomial series has a logical meaning, since he was dissatisfied with the proof he found in his textbook. He was a rigorist from the very beginning and the impulse he received when he worked with the binomial theorem as a boy was to be developed further in several of his best works.

Arithmetical Investigations.

The number-theoretical ideas which streamed from Gauss during the fruitful years 1795–1801 were for the most part collected together in the work that he published in Leipzig in 1801, *Disquisitiones arithmeticae*—Arithmetical investigations. It is Gauss's *magnum opus*, which in one stroke made number theory a firmly grounded and coherent part of mathematics.

The cost of printing, as was the case of his doctor's thesis, was paid by Duke Ferdinand. The work opens with a dedication to "His most Graceous Highness, Prince and Lord Carl Wilhelm Ferdinand, Duke of Braunschweig and Lüneburg." In it Gauss says, among other things, that without the Duke's goodness "I should never have been able completely to devote myself to mathematics, to which I have always been drawn with passionate love." The dedication is supplied in the rococo style demanded by the customs of the period, but in this case the homage was not empty flattery. For these words reflected Gauss's real feelings. In the rhetoric of the time, one may say that Duke Ferdinand, who met his fate and his death in the battle against Napoleon's army at Jena and Auerstädt, was handed a crown of victory that will never wither, in the *Disquisitiones* of the twenty-four-year-old Gauss.

Before proceeding, let us say a few words about the language of the work. Like Gauss's work of earlier years, it is written in Latin, which was then still the international language of science. Under the influence of nationalist feelings at the beginning of the nineteenth century, however, Gauss later changed over and began to write in German. If he and other researchers had stood their ground with Latin, perhaps we could have avoided the present-day confusion of languages. Perhaps we would still have a reasonably easy to understand "scientific Latin" which could be read by all scientists on earth. Present-day attempts with different international languages, such as Interlingua, are praiseworthy, but it is always the first step that counts, and the beginning hurdles which the new languages always have such difficulty in surmounting need not have arisen if we had simply stayed with Latin.

Arithmetical investigations is divided into seven parts:

Congruences in general
Congruences of the first degree
Residues of powers
Congruences of the second degree

Quadratic forms
Applications
Division of the circle

The work has generally been judged extremely hard to read. The distinguished German mathematician Peter Gustav Dirichlet (1805–1859) who later became Gauss's successor at Göttingen, was the first to disseminate their contents in his lectures. We shall here set down only a few samples of Gauss's exploits in number theory, illustrating these results with some examples.

Congruences in General and Congruences of First Degree.

On the first page Gauss introduced a new mathematical symbol, one of his revolutionary improvements in the nomenclature of number theory, which simplifies immensely the handling of the notion of arithmetic divisibility. (This chapter is concerned solely with whole numbers, that is to say, the numbers $0, \pm 1, \pm 2, \pm 3, \ldots$.)

If the number m divides the difference $a - b$ (or $b - a$) of two numbers a and b without remainder, then a and b are said to be *congruent modulo m*, and Gauss wrote

$$a \equiv b \pmod{m}.$$

This expression is read: *a is congruent to b modulo m*. The relation is called a *congruence*; and m is called the *modulus* of the congruence. The number b is called the *residue* of a modulo m, and conversely a is called a residue of b modulo m. If the difference $a - b$ is not divisible by m, then a and b are said to be *incongruent modulo m*, and a and b are said to be *nonresidues* of each other, modulo m. We have for example

$$25 \equiv 10 \ (\text{mod } 3) \text{ and } -7 \equiv 15 \ (\text{mod } 11),$$

since $25 - 10 = 15$ is divisible by 3, and $-7 - 15 = -22$ is divisible by 11. The number 10 is a residue of 25 with respect to the modulus 3, while 10 is a nonresidue of 7 with respect to the modulus 11.

According to the definition, $a \equiv b \ (\text{mod } m)$ means the same as $a - b = m \times y$, where y is some whole number. If for the given numbers a, b, and m there is no number y that fulfills this condition, then b is a nonresidue of a modulo m. The congruence $a \equiv 0 \ (\text{mod } m)$ means that a is divisible by m, for example $35 \equiv 0 \ (\text{mod } 7)$.

Gauss chose the symbol $\equiv$ (which reminds us of the symbol $=$) with great foresight, since there is close analogy between congruences and equalities. The notion of congruence is however more inclusive, since one may interpret an equality as a congruence with modulus 0.

Congruences have purely practical applications, as for example in train schedules, school calendars, or timetables in general. In a mathematical contest for schoolchildren in the newspaper "Svenska Dagbladet" of 17 October, 1962, this problem appeared, among others: "Today is Wednesday the 17th of October 1962. On which day of the week will the 17th of October fall in the year 3000? Leap year occurs when the number of the year is devisible by four, with the exception that years which are divisible by 100 are leap years only if the number is divisible by 400."

We urge the reader to solve this problem on his own. It is clearly a question of a congruence property (mod 7). The calculation of the day of the week is complicated by leap years. A solution that does not use congruences directly is to be found at the end of this chapter.

In questions about addition, subtraction, and multiplication of congruences with the same modulus, there are arithmetical laws analogous to those for equalities. For example, we shall show that the Fermat number $2^{2^5}+1 = 2^{32}+1$ which we mentioned earlier has the factor 641 (cf. p. 35).

First we calculate with equalities, then with congruences. We start with the equalities.

$$5 \times 2^7 = 641 - 1, \tag{1}$$

$$5^4 = 641 - 2^4. \tag{2}$$

If we raise (1) to the fourth power we have

$5^4 \times 2^{28} = (641 - 1)^4 = 641^4 - 4 \times 641^3 + 6 \times 641^2 - 4 \times 641 + 1$, so that $5^4 \times 2^{28} - 1 = 641 \ (641^3 - 4 \times 641^2 + 6 \times 641 - 4)$,

or, if we denote the number inside the parentheses by k,

$$5^4 \times 2^{28} - 1 = 641 \times k, \tag{3}$$

where k is a natural number. If we combine (2) and (3) we obtain

$$(641 - 2^4) \times 2^{28} - 1 = 641 \times k,$$
$$641 \times 2^{28} - 2^{32} - 1 = 641 \times k,$$

and

$$2^{32} + 1 = 641(2^{28} - k) = 641m, \tag{4}$$

where m is a natural number. From (4) it follows directly that 641 is a factor of $2^{32} + 1$.

In congruence notation, (1) is written

$$5 \times 2^7 \equiv -1 \ (\text{mod } 641);$$

and by raising this to the fourth power we obtain (3) as a congruence. That is to say,

$$5^4 \times 2^{28} \equiv 1 \; (\mathrm{mod} \; 641). \tag{5}$$

Expression (2) becomes the congruence

$$5^4 \equiv -2^4 \; (\mathrm{mod} \; 641),$$

and in combination with (5) it becomes

$$-2^{32} \equiv 1 \; (\mathrm{mod} \; 641) \; \text{or} \; 2^{32}+1 \equiv 0 \; (\mathrm{mod} \; 641);$$

in other words, $2^{32}+1$ is divisible by 641.

Our second example illustrates the well-known criteria for dividing a number by 9 or 11. In the decimal system an arbitrary number N can be written as

$$N = 10^k \times a_k + 10^{k-1} \times a_{k-1} + \ldots + 10 \times a_1 + a_0,$$

where $a_0, a_1, \ldots, a_{k-1}, a_k$ are chosen from among the numbers 0, 1, 2, ..., 9.

The rule for 9 is: the number N can be divided by 9 only if the sum of the digits $a_0 + a_1 + \ldots + a_{k-1} + a_k$ is divisible by 9. The assertion follows from the fact that from the congruence $10 \equiv 1 \; (\mathrm{mod} \; 9)$ we have $10^m \equiv 1 \; (\mathrm{mod} \; 9)$, for $m = 1, 2, \ldots, k$.

The rule for 11 is: the number N can be divided by 11 only if $a_0 - a_1 + a_2 - \ldots + (-1)^k a_k$ is divisible by 11. It follows from the fact that the congruence $10 \equiv -1 \; (\mathrm{mod} \; 11)$ implies $10^m \equiv (-1)^m \; (\mathrm{mod} \; 11)$, for $m = 1, 2, \ldots, k$. As a check the reader can carry out the proof with equalities. In so doing the advantage of congruences will be made apparent.

In the second section of his Arithmetical Investigations Gauss first proves several theorems from which emerge what is usually called *the fundamental theorem of arithmetic: Every natural number larger than 1 can, except for the order of the factors, be written in only one way as the product of prime numbers.*

Examples: $130 = 2 \times 5 \times 13$; $250 = 2 \times 5 \times 5 \times 5$.

This theorem, like the fundamental theorem of algebra, appears more or less obvious, but it is just such obvious theorems that are often difficult to prove, because circular proofs lie near at hand: more or less unconsciously we assume the truth of what we wish to prove.

By using the fundamental theorem Gauss then determined the *greatest common divisor* (a,b) and the *least common multiple* $\{a,b\}$ of two numbers a and b. (Naturally one can also consider more than two numbers.) For example, if $a = 24 = 2^3 \times 3$ and $b = 90 = 2 \times 3^2 \times 5$, we have $(24,90) = 2 \times 3 = 6$ and $\{24,90\} = 2^3 \times 3^2 \times 5 = 360$.

If a and b are natural numbers, then

$$a \times b = (a,b) \times \{a,b\}.$$

We can verify this in our example, where we find that

$$24 \times 90 = 6 \times 360 = 2{,}160.$$

If the greatest common divisor (a,b) of two whole numbers is 1, then a and b are said to be *relatively prime*. We see that $(6,35) = 1$, for example.

In analogy with algebraic equations of the first degree, Gauss then went on to congruences of the first degree

$$ax \equiv b \;(\text{mod } m).$$

This expression is equivalent to the equality

$$ax - my = b,$$

where a, b and m are given whole numbers and x and y are unknown whole numbers. Such equations are called Diophantine equations after the Alexandrian Greek Diophantos (ca. A.D. 250–330).

Gauss's result for the solvability of linear congruences is:

If $(a,m) = d$, then a necessary and sufficient condition for the congruence

$$ax \equiv b \;(\text{mod } m)$$

to be solvable is that d be a divisor of b. There then exist d different sequences of solutions, or, more succinctly, d solutions. We shall give several examples.

Example 1. The congruence $x \equiv 4 \;(\text{mod } 7)$ needs little analysis, since we see immediately that the only solutions are the numbers $x = 4 + 7k$, where k is a whole number.

Example 2. The congruence $6x \equiv 5 \;(\text{mod } 8)$ is not solvable, because $d = (6,8) = 2$ and 2 is not a divisor of 5.

Example 3. The congruence $2x \equiv 3 \;(\text{mod } 5)$ is solvable, since $d = (2,5) = 1$; that is to say, the numbers 2 and 5 are relatively prime, and we have one solution. Through substitution we find that $x = 4$ answers the requirements. The Diophantine equation $2x - 5y = 3$ gives $x = 4 + 5k$, where k is an arbitrary whole number. We write this in convenient notation as $x \equiv 4 \;(\text{mod } 5)$.

Example 4. The congruence $34x \equiv 60 \;(\text{mod } 98)$ is solvable, since $d = (34,98) = 2$ and 2 is a divisor of 60. Thus there exist *two* solutions: $x \equiv -4 \;(\text{mod } 98)$ and $x \equiv 45 \;(\text{mod } 98)$. We may write $x \equiv -4, 45 \;(\text{mod } 98)$.

Congruences of Second Degree.

In the third and fourth sections Gauss continued on to congruences of higher degree. Especially important is the binomial congruence $x^n \equiv 1 \;(\text{mod } m)$, of which a simple example is

$$x^2 \equiv 1 \;(\text{mod } 8).$$

It has four solutions:

$$x \equiv 1, 3, 5, 7 \;(\text{mod } 8).$$

Another example is the result known as the *little theorem of Fermat*, which was stated by Fermat, but probably known much earlier by the

Hindus and Chinese. Gauss formulated and proved it with congruences. It runs as follows:

   If p is a prime number and a is a whole number that is not divisible by p, then

   $a^{p-1} \equiv 1 \pmod{p}$

Consider two examples. For $p = 7$ and $a = 2$ we have $2^6 \equiv 1 \pmod 7$; that is to say, $2^6 - 1 = 63$ is divisible by 7, an obvious fact. For $p = 37$ and $a = 10$ we find that the number $10^{36} - 1$ is divisible by 37; this we can check by calculation.

   The fourth section deals with one of the most interesting parts of number theory, the *theory of quadratic residues*. A number a is called a quadratic residue of the number m if the congruence

   $x^2 \equiv a \pmod{m}$

has a solution. If the congruence has no solution, then a is called a quadratic nonresidue of m.

   These names were not introduced by Gauss but by Euler, who had worked earlier with the question of determining whether a number is a quadratic residue or nonresidue of another number.

   For example, is 3 a quadratic residue of 11? If so, then we should be able to solve the congruence

   $x^2 \equiv 3 \pmod{11}$.

If we experiment with $x = 1, 2, 3$ and so forth, we find that $x = 5$ is a solution. Thus 3 is a quadratic residue of 11. On the other hand, the congruence

   $x^2 \equiv 6 \pmod{11}$

is not solvable: 6 is a quadratic nonresidue of 11.

In the general investigation one may restrict one's attention to the case in which a and m are (different) prime numbers. Then the question becomes: *If p is a given prime number, for which prime numbers q can we solve*

$$x^2 \equiv q \pmod{p}?$$

In this general form Gauss could not solve the question—nor has anyone else been able to. But there is a curious relationship between this congruence and the congruence

$$x^2 \equiv p \pmod{q}.$$

This relation had been set forth in turn by Euler and Legendre (the latter called it the *law of quadratic reciprocity*), but neither had been able to give a rigorous proof for what they had discovered by numerical calculations. As mentioned before, Gauss discovered this theorem on his own in 1795 and published the first correct proof. He called it the *fundamental theorem in the theory of quadratic residues*. It runs as follows:

For the pair of congruences

$$x^2 \equiv q \pmod{p} \text{ and } x^2 \equiv p \pmod{q},$$

where p and q are different primes, the following holds: either both congruences are solvable or both are unsolvable, except in the case in which both p and q have remainder 3 when divided by 4, in which case one of the congruences is solvable while the other is unsolvable.

The foundation of Gauss's proof of this theorem was his experimentation with numbers, which certainly is unique in the history of mathematics. Of course we are speaking here of people and not of machines. During the year 1795 and earlier, while Gauss still did not know that he had predecessors in this area, he made up a huge table of prime numbers; of quadratic residues and nonresidues; and of fractions $1/n$ from $n = 1$ to $n = 1,000$, expressed as periodic decimals with the entire period given.

The longest period that the decimal expansion of the fraction $1/n$

can have is $(n-1)$ digits. For example,

$$\frac{1}{7} = 0.142857\ 142857\ldots,$$

has the maximal length of period, 6 digits, while

$$\frac{1}{9} = 0.11111\ldots$$

has only one digit in its period.

In computing the entire period of the decimal expansions of $1/n$ for $n = 1$ up to $n = 1,000$, in a good many cases Gauss had to calculate several hundred decimals. For example, he determined $1/811$ to 822 decimals, the last few being put in as a check of the calculations. Gauss wrote up this table to find the connection between the period of the decimal expression and the denominator n. It was a frightfully laborious path that he set out upon, but it finally led him to his goal.

On April 8, 1796, a short notice in the journal states that he had found the first exact proof of the fundamental theorem of quadratic residues. It was a very long proof, which contained eight different cases and was carried out with obstinate logic. The great German mathematician Leopold Kronecker (1823–1891) later characterized it as the "test of strength of Gauss's genius."

True to his principle of the finished work of art, Gauss worked on and presented a total of eight proofs of "the gem of arithmetic," which really deserves its name; it plays a fundamental role in higher number theory and in several areas of algebra.

The law of quadratic reciprocity can be formulated in various ways. The shortest is probably the following:

A prime number p is a quadratic residue or nonresidue of another prime number q according to whether

$$q \times (-1)^{(q-1)/2}$$

is a residue or nonresidue of p.

We shall illustrate the different cases that can occur (in the first formulation) with a few examples.

Example 1. $p = 23$, $q = 13$. Note that 23 yields the remainder 3 while 13 yields the remainder 1 when divided by 4. Thus the two congruences

$$x^2 \equiv 13 \pmod{23} \text{ and } x^2 \equiv 23 \pmod{13}$$

are either both solvable or both unsolvable. They are solvable. For example, the first one has the solution $x = 6$, and the second one also has the solution $x = 6$.

Example 2. $p = 5$, $q = 17$. Since both of these numbers have remainder 1 when divided by 4, it must be true that either both of the congruences

$$x^2 \equiv 17 \pmod{5} \text{ and } x^2 \equiv 5 \pmod{17}$$

are solvable or both are unsolvable. They are unsolvable.

Example 3. $p = 23$, $q = 11$. In this case both numbers have remainder 3 when divided by 4. Of the two congruences

$$x^2 \equiv 11 \pmod{23} \text{ and } x^2 \equiv 23 \pmod{11},$$

one must be solvable while the other is unsolvable. The first is unsolvable. The second has the solutions 10, 21, 32, and 43, for example.

Quadratic Forms.

In the fifth section Gauss first handles *binary quadratic forms*, that is, expressions of the form

$$ax^2 + 2bxy + cy^2.$$

Here the problem is to determine the whole-number solutions in x and y of the Diothantine equation

$$ax^2 + 2bxy + cy^2 = m,$$

where a, b, c, and m are given whole numbers. He then studied the corresponding problem for *ternary quadratic forms*, that is, expressions of the type

$$ax^2 + by^2 + cz^2 + 2dxy + 2exz + 2fyz.$$

In the sixth section the foregoing theory is applied to a number of special cases, such as to whole-number solutions of the diophantine equation

$$ax^2 + by^2 = m.$$

Dividing the Circle.

In the seventh and final section Gauss applies his earlier results to the binomial congruence $x^n \equiv 1 \pmod{p}$, where p is a prime number and n a natural number. The relation between these arithmetic congruences and the binomial equation $x^n = 1$ gives the solution to the problem of dividing the circle and constructing the regular 17-gon of which we spoke earlier. The binomial congruence $x^n \equiv 1 \pmod{p}$ unites arithmetic, algebra, and geometry in one of the great syntheses which Gauss pursued and which he achieved here in a fashion that has few counterparts in the history of mathematics. "*Disquisitiones arithmeticae*" is a classical symphony in seven movements, where the different *leitmotifs* are combined in a finale that is carried out with strength and clarity.

Solution of the Day of the Week Problem.

We return to a problem that we stated earlier, namely, to determine the number of days from October 17, 1962 (exclusive) to October 17, 3000 (inclusive), and to find the remainder this number yields when divided by 7 (the number of days in a week). The number of days we are after comprise a number of entire years, namely $3000 - 1962 = 1038$. Starting with the year 1964 (the first year divisible by 4), up to and including the year 2999 there are $1036/4 = 259$ years that are divisible by 4. Of these there are 10 that are divisible by 100, while 3 years (2000, 2400, and 2800) are divisible by 400. The number of leap years (3000 is *not* a leap year) is thus $259 - 10 + 3 = 252$. The number of days x that we seek is thus

$$x = 365 \times 1038 + 252 = (7 \times 52 + 1) \times 1038 + 7 \times 36 = 7 \times 52 \times 1038$$
$$+ 1038 + 7 \times 36 = 7 \times 52 \times 1038 + 7 \times 148 + 2 + 7 \times 36 = 7 \times (52 \times 1038$$
$$+ 148 + 36) + 2.$$

This is a number of entire weeks plus 2 days, which implies that October 17, 3000 is a Friday. We may write $x \equiv 2 \pmod{7}$.

One can use congruences for solving many different calendar problems although the formulas become forbiddingly complicated. Perhaps the most famous is Gauss's Easter formula for determining the date of Easter Sunday in any arbitrary year. It is valid for the period 1700–1899. Gauss claimed that he was motivated to produce it because his mother did not know the exact date on which he was born; she only knew that he was born on a Wednesday, eight days after Ascension Day.

FOUR ASTRONOMY

The Determination of Ceres' Orbit.

Arithmetical investigations made Gauss famous among mathematicians. The determination of the orbit of the planetoid Ceres made him famous in all of the academic circles around the world.

On January 1, 1801—a half-year before *Arithmetical investigations* appeared—the Italian astronomer Joseph Piazzi in the observatory at Palermo had discovered a moving object of the eighth magnitude in the constellation of Aries; at first the experts disagreed as to whether the new heavenly body was a planet or a comet.

The question also had a philosophical background. About that time the great Hegel published an article in which he spoke ironically of the astronomers' search for an eighth planet; he thought he was able to prove by purely logical argument that the number of planets was exactly seven, neither more or less!

Piazzi's observations extended over a period of forty-one days, during which the orbit swept out an angle of only nine degrees. After that the object was lost to sight—its light vanished in the rays of the sun—and neither Piazzi nor any other astronomer could find it again. Then the problem was to use these meager observations as a guide in calculating the orbit of the vanished celestial body, so that observers would know where to point their telescopes.

It was a new question. When Herschel made his famous discovery of the planet Uranus in 1781 it had been sufficient to assume that the orbit was circular, and when one calculated the orbit of comets one could assume they were parabolas.

The general form of a planet's or comet's orbit depends upon its eccentricity, which is a measurement of the orbit's deviation from circularity. The circle's eccentricity is zero while the parabola's eccentricity is one. But in the case of this new planet one could only assume that its orbit was an ellipse, that is to say, that its eccentricity was an unknown number between zero and one. To be sure, this case had been dealt with by Euler, Lambert, Lagrange, and Laplace among others, but they had used methods of great difficulty, which did not allow a complete determination of the orbit from observations that spanned only a short period of time. Even Pierre-Simon Laplace (1749–1827), a master of celestial mechanics, thought that the problem was unsolvable in this form. Thus in the case of Ceres the known methods were useless.

Gauss had already worked with astronomical questions, e.g., with the theory of the motion of the moon. He was now attracted to this new problem, which looked as though it would give free play to his overpowering combination of computational virtuosity and creative imagination. He decided to work out more useful methods for determining orbits, and was soon ready with his first solution. Thanks to his ephemeris (an astronomical almanac in which the daily positions of the heavenly bodies are listed in advance) Ceres was found again during the period November 25 to December 31, 1801 at almost exactly the points he predicted. The first to find it again was Zach in Gotha on December 7, and then Olbers in Bremen on New Year's day of 1802.

Wilhelm Olbers (1758–1840) was one of the foremost amateur astronomers of all time. During the day a practicing physician, at night he sat in his private observatory, where among other things he discovered no fewer than six comets and two planetoids. This man, who clearly needed very little sleep, became one of Gauss's closest friends.

To Hegel's growing displeasure, people began to discover one new planet after another. These were the asteroids, small planets forming a belt in the great gap between Mars and Jupiter. In 1802 Olbers found the planetoid Pallas in the immediate vicinity of Ceres, in 1804 Harding found Juno, in 1807 Olbers found Vesta, and so on. Today more than 1,500 planetoids have been identified. Ceres, with its diameter of 765 kilometers, is still the largest. From there they fall off to lumps of rock that are only one to two kilometers in diameter.

While continually improving his methods, Gauss calculated ephemerides for the new planetoids as they were discovered. His calculations for Vesta provoked Olber's unconcealed astonishment. In only ten hours Gauss had calculated the elements of the orbit and in addition to that compared his theoretical values with different observations of the new planetoid.

When it was a matter of parabolic orbits his calculations were even faster. Gauss could calculate the orbit of a comet in a single hour; it had taken Euler, using the older methods, three days. Such a calculation is said to have made Euler blind in one eye. Gauss asserted somewhat heartlessly: "I should also have gone blind if I had calculated in that fashion for three days."

The Theory of Motion of Celestial Bodies.

Gauss published his new methods in 1809 as *Theoria motus corporum coelestium in sectionibus conicis solem ambientium*—Theory of the motion of the heavenly bodies moving about the sun in conic sections. (The curves that result from the intersection of a cone and a plane were called *conic sections* by the Greek geometers. Among them are circles, ellipses, and parabolas.) It was first written in German, but at the request of the well-known publisher Perthes, Gauss changed it to Latin. He wrote to Perthes that he was convinced that *Theoria motus* would be studied for centuries.

His convictions were correct. He was conscious of his ability and feigned no false modesty. In the world of reality he was the counterpart of the fictional genius Sherlock Holmes. In the novel "The Greek Interpreter," Sherlock says: "My dear Watson, I cannot agree with those who rank modesty among the virtues. To the logician all things should be seen exactly as they are, and to underestimate one's self is as much a departure from truth as to exaggerate one's own powers."

Ordinary mortals must, like Doctor Watson, grumble a reluctant agreement. On another occasion Gauss said that his calculation of Ceres' orbit was only a confirmation of Newton's law of gravity. That too was correct.

Theoria motus is thus a classical work in theoretical astronomy. Gauss's method of determining an elliptical orbit from three complete observations was later improved with respect to its practical use, primarily by Gauss's student J. F. Encke, and during the past few years it has been adapted to modern data-processing machines. But the fundamental principles are still those of Gauss, and nothing better has ever been brought forward.

The conic section for a planet's orbit is derived under the initial assumption that the planet is acted upon by *one* central force, which acts from the center of the sun. Thus force is calculated using Newton's law of gravity. It is a "two-body" problem. In reality the problem is much more complicated, since the other bodies in the solar system also influence the planet's orbit through their attraction. In its general form this problem still has no *exact* mathematical solution. No one has been able to solve it even for three bodies—the famous "three-body" problem—except in certain special cases.

But the problem can be solved *approximately* in each given case. The planets' distances from one another are large, and their masses are small with respect to that of the sun. Thus as a rule the deviation from an elliptical orbit is small and is not noticed, for example, during a single revolution. For longer intervals of time however, perturbations from the other bodies in the solar system must be taken into account. In the mathematical treatment of this problem one uses series expansions of different kinds, which approximate the exact solutions.

Perturbation theory was founded by Newton and later was developed principally by Laplace. It is far from simple. The orbit of the moon is said to be the only problem to cause Newton a headache, and Laplace, who could put to good use the enormous development of mathematics since Newton's time, needed the years from 1799 to 1825 to produce the five volumes of his magnificent *Treatise on celestial mechanics, Traité de mécanique céleste*.

True to his principle that a work should have such a degree of completion "that nothing more could be desired"—*ut nihil amplius desiderari possit*—Gauss published very little of his contributions to perturbation theory. But he communicated them to his students; for example the orbit computed for the comet named after Encke was computed by Gauss's method.

Only when Gauss's posthumous papers were published in his Collected Works did one get a clear idea of his fundamental contributions to perturbation theory through his calculations for Ceres and Pallas. In his first calculations of the perturbations of Ceres, Gauss used his own research on the hypergeometric series and the arithmetic-geometric mean, and also made use of Laplace's work in an essential way.

The calculations of the perturbations of Pallas within the bounds of accuracy of the observations consisted mostly of page after page of numerical calculations with no explanatory text. This difficult puzzle was put into order and published by the German astronomer Martin Brendel, who in his commentary points out among other things that Gauss has almost completed an enormous problem, "which even today an astronomer does not gladly set out upon."

But even Gauss could tire of mechanical calculations. He used to say there is poetry in a table of logarithms, and he certainly enjoyed his incredible numerical skill, but he certainly did not pursue it as an end in itself.

Concurrently with his calculations concerning Pallas, Gauss was working in purely theoretical research concerning the secular perturbations of a planet's orbit. In contrast to the periodic perturbations, which are repeated within a relatively short period of time, the secular perturbations grow (or decline) slowly and steadily. During astronomical periods of time their influence can be decisive in determining the stability of the solar system, the stability that Laplace thought he had proved by using the mathematical theorems of his celestial mechanics, the stability that Newton almost certainly thought God's finger must influence in order to keep the planetary system, with all of its screws coming loose, from being thrown out into the Milky Way.

Gauss made no statement about stability theory, but in a paper dated 1818 he calculated theoretically the secular perturbations of the first degree of orbital elements, to which a planet is subjected by other planets. A long Latin title shows how Gauss pictured a model of the secular perturbations propagated by a planet. His conception was that the planet's mass could be thought of as distributed over its entire orbit, with the distribution made in such a way that the mass at every sector of the elliptical ring is inversely proportional to the velocity of the planet in that sector of the orbit. The attraction that this elliptical ring exercises over another body is exactly equal to its secular perturbations. This model is characteristic of Gauss's way of thinking. The perturbation problem is not just a mass of formulas and figures which he handled with his usual virtuosity, but also a living picture which gives clarity to the mathematical symbols. From a purely mathematical viewpoint this treatise was surprising in that Gauss carried out his solutions with the help of what are called *elliptic integrals*.

Gauss was also interested in purely practical astronomy. In 1801 he procured his own sextant, and for fifty years he shirked neither long and laborious observations nor long and laborious calculations—although he naturally was aware that he could have put his time to better use. The last observation which he carried out himself and communicated to the journal "Astronomische Nachrichten" was an eclipse of the sun on July 28, 1851. He was then 74 years old.

In this connection one cannot pass up the opportunity to speculate over the amount of time Gauss would have saved had he had access to modern computers for his numerical calculations and to the automatic methods for observations of our time. Would he have used the time to carry out and complete all of the mathematical ideas in his journal? It seems reasonable that the answer is "yes." But it is not an unconditional "yes," since it is not probable that Gauss, with his universal outlook, would have used his spare time only for pure mathematics. In a letter of 1803 to Bolyai he writes: "Astronomy and pure mathematics are now once and for all the magnetic poles to which my soul's compass shall always point."

Gauss is Called to a Professorship. His First Marriage.

The first sign of Gauss's international fame came from St. Petersburg (now Leningrad), when he was elected to a corresponding membership in its Academy of Science on January 31, 1802, less than two months after the rediscovery of Ceres. In September of the same year he was offered a post as director of the astronomical observatory in St. Petersburg. In the beginning of 1803 Gauss decided to stay in Braunschweig. Duke Ferdinand had increased his support, but this certainly was not the sole reason for Gauss's decision. In addition to being loyal to his patron, he was loyal to his native land. Gauss was of a rooted nature and during his entire life never traveled far from Göttingen.

In spite of the Duke's increased support, Gauss's economic position was insecure, since he had no permanent employment. He was considered for a post at the Collegium Carolinum in Braunschweig, but on the basis of pedagogical merit it went to his former teacher Hellwig.

In 1804 a new observatory was planned in Göttingen with Gauss as director; the University authorities did not want to miss the chance of adding his luster to the institution. Gauss took the waiting period calmly. He lived most of the time in his home town, but at the house at Steinweg 22 rather than with his parents; once in a while he visited in Göttingen, and once in a while he visited Olbers in Bremen or Zach in Gotha.

Through Olbers, Gauss came in contact with Friedrich Wilhelm Bessel (1784–1846), who at the time was employed by a business firm in Bremen. He subsequently became both a practical and theoretical astronomer of the first rank. Much later Gauss said: "Olbers performed great, very great services for astronomy, but his greatest service lies in the fact that he immediately recognized Bessel's gift for astronomy, and that he won and fostered it for science." In the light of Gauss's close friendship with Olbers, this observation is of Olympian objectivity—and of Olympian coldness, from the icy spaces where such judgments crystallize.

On July 25, 1807, Gauss was named Professor of Astronomy and director of the new observatory in Göttingen. Juno's discoverer, Carl Ludwig Harding, was named to the position of Observer at the same time, and in 1812 he also became a Professor of Astronomy. It is not surprising that Gauss, who was really a mathematician, became a Professor of Astronomy, if one considers the fact that at that time there was scarcely any differentiation between "pure" and "applied" mathematics, and that Gauss worked with equal mastery in both fields. Even today the boundary between pure and applied mathematics is blurred. The division of mathematics into different subjects is certainly advanced by the growing body of knowledge that leads perforce to specialization.

In November of 1807 Gauss moved to Göttingen, where he remained in his official position for the rest of his life. But he did not come alone.

In 1803 he had met Johanna Osthoff, the daughter of a tannery owner in Braunschweig. She was born in 1780 and was an only child. Her lovable and good-hearted nature gave Gauss a much needed distraction from celestial mechanics. After having called upon her for a year, Gauss wrote a romantic proposal of marriage. Johanna delayed her affirmative answer for three months, not because she doubted her own feelings, but probably because of the shyness that was expected in those days; perhaps she felt it would be difficult to fill the gulf between her own and her future husband's world of thought. They were engaged at the end of 1804 and immediately afterward Gauss wrote an exuberant letter to Bolyai, in which, after first having mentioned and commented upon the discovery of the planetoid Juno, he said, among other things: "For three days now this angel, almost too heavenly for our earth, has been my fiancée. . . . Life lies before me like an eternal spring with new radiant colors."

Johanna's delicate and charming disposition is witnessed by all who knew her, but unfortunately no picture of her is preserved. Gauss himself had a Nordic appearance. His hair was blond and his eyes were dark blue and of penetrating clarity. He had a prominent nose and was somewhat less than average in height.

The wedding took place on October 9, 1805. The couple lived on in Braunschweig for a time, in the house which Gauss had occupied as a bachelor. Their private joy was unclouded, and their economic position improved when the Duke—probably influenced by the fact that Gauss was again invited to come to St. Petersburg—once again increased Gauss's stipend. But foreign events were moving in unfortunate directions for the Duke of Braunschweig. When Gauss visited him in May of 1806 to thank him for the latest increase in salary it was the last time he met his benefactor, who was of a certainty also his friend.

The Death of the Duke. The Death
of Gauss's Wife. His Second Marriage.

On October 14 the battle of Auerstädt, in which Duke Ferdinand was commander of the Prussian and Saxon troops, took place. In it the Duke was struck by a musket ball that entered above his right eye and carried away his left eye. Napoleon showed little magnanimity toward his defeated and sorely wounded enemy and scornfully turned away a deputation that pleaded that the seventy-one year old Duke be allowed to die in peace among his people.

The French army was just at the point of marching into Braunschweig, and flight became the Duke's only course of action. From his window at Steinweg 22, across the street from the palace gates, Gauss saw the Duke's carriage travel away one morning toward the end of October, on the road to Altona. Duke Ferdinand died at Altona on November 10, 1806. Perhaps his name does not live in the history of war, but it lives on in mathematics.

The Duke's tragic death made a deep impression on Gauss. He

was of a serious and hypocondriachal nature, and after his friend's death he regularly succumed to variable periods of anxiety, bordering on melancholia.

During this period of political upheavals Gauss found compensation in his work and his family. On August 21, 1806, his first son Joseph was born; he received his name after Piazzi, the discoverer of Ceres. (Four of Gauss's six children were christened in that way, after discoverers of planetoids.) In May of 1808 Gauss wrote to Bolyai that little Joseph was everyone's darling, but he then pointed out: "A mathematician is scarcely in him, he is too wild, too playful, I would say."

On February 29, 1808 a daughter followed, and Gauss jokingly complained that she would only have a birthday every fourth year. As a mark of respect to Olbers she was christened Wilhelmina (Minchen or Minna). The third child, a son, born on September 10, 1809, was named Ludwig, after Harding, but was called Louis.

The second childbirth had drained the mother's strength. The third was too much. After a difficult delivery, Johanna died on October 11, 1809, and she was followed five months later by Louis— *der arme kleine Louis* as his father always sadly said. He died suddenly on March 1, 1810.

The change of fortune was as abrupt as a bolt of lightning. In September of 1808 Gauss had written to Bolyai: "The days fly happily by in home life's daily course; when our daughter gets a new tooth or our son learns a couple of new words, it is almost as important as when a new star or a new truth is discovered." At the end of October in 1809 he wrote—probably while staying with Olbers—a lamentation over his wife's death, on two sheets of tear-stained folio paper. " 'I should not yield too much to sorrow,' these were nearly your last words. O, how shall I begin to shake it off? O, beseech the Eternal, could He deny you anything?—only this one thing, that your infinite goodness, always living, should be recollected to me, so that I, as much as a poor son of the earth is able, can be as you were,"

The instinct for self-preservation and his growing vitality were strong enough to lift Gauss from his slough of depression. It came about with unusual speed. It is a well-known psychological or biological fact that one who has been happy in his marriage often remarries with surprising speed after his wife's death. Gauss belongs to that category.

On March 27, 1810 he wrote his second proposal of marriage. He proposed to his late wife's closest friend of her last years, Minna Waldeck. Born in 1788, she was the youngest daughter of a Professor of Law, Johann Peter Waldeck, of Göttingen. Gauss married her on August 4, 1810. The new marriage was a happy solution to Gauss's nonscientific problems. Minna handled the motherless children with as much tenderness as if they had been her own, and even if Gauss—as he wrote in his proposal—could offer her only "a divided heart in which the picture of the deceased should never be erased," Minna still won his sincere love.

Two sons and a daughter were born in the new marriage, Eugene on July 29, 1811, Wilhelm on October 23, 1813, and Therese on June 9, 1816.

After his arrival at Göttingen in 1807, Gauss worked at first in the old observatory. It took a long time before the new one was ready. The period was not conducive to peaceful work. In accordance with the Treaty of Tilsit of 1807 Prussia had to cede a large area to France. Göttingen came to lie in the new kingdom of Westphalia, and the French exactions were severe. They even fell upon Gauss. As a professor at Göttingen it was his duty to pay 2,000 francs, a large sum in that day's currency, which Gauss had no possibility of producing.

Various friends wished to help him, but Gauss had clearly wearied of financial assistance. Olbers, who had a private fortune, made the first attempt, but Gauss thanked him for his benevolence and returned the sum. Then Laplace wrote a letter from Paris in which he informed Gauss that he considered it an honor to pay the sum and lift an undeserved burden from a friend's shoulders. An anonymous person then sent 1,000 guilders to Gauss, who accepted the gift since he could not find out who the sender was. But he repaid Laplace with interest at the current rate of exchange.

Gauss harbored an understandable animosity toward Napoleon and the French for what they had done to his land and to his great patron. But this animosity did not hold for Laplace, whom he placed in a class after Newton. On his side Laplace recognized Gauss's greatness without reservation. The famous explorer Alexander von Humboldt (1769–1859) was one of Gauss's supporters for the directorship of the observatory in Göttingen. But he wanted to have an expert expression of opinion and therefore asked Laplace to tell him the name of Germany's greatest mathematician: "Pfaff," answered Laplace. "But what do you think about Gauss?" wondered the dumbfounded von Humboldt. "Gauss is the world's greatest mathematician", answered Laplace in the same laconic way.

By 1814 the exterior of the new observatory was ready. The professors' residences were completed in 1816, and Gauss and his family moved into the west wing, while Harding lived in the east. During the following years, Gauss and Harding installed the astronomical instruments. New ones were ordered in Munich. Among other times, Gauss visited Munich in 1816, but it was not until 1821 that the instruments arrived at Göttingen.

After the intense sorrow of Johanna's death had been mollified in his second marriage, Gauss lived an ordinary academic life which was hardly disturbed by the violent events of the time. His powers and his productivity were unimpaired, and he continued with a work program which in a short time would have brought an ordinary man to collapse.

From the old observatory he saw The Great Comet suddenly appear in the sky in August of 1811, and calculated its parabolic orbit with his usual precision. For the people of Europe it was an omen that portended the burning of Moscow and the defeat of *la Grande Armée*.

SIX

OBSERVATIONAL ERRORS AND THE CALCULUS OF PROBABILITIES

Gauss's great interest in astronomy, and his later interest in geodesy, compelled him to seek rational methods for determining the magnitude of observational errors. In turn, the theory of observational errors forced him to deal with the modes of thought and concepts of the calculus of probabilities. This work had great significance in the development of numerous areas in both the calculus of probabilities and mathematical statistics. For example, the theory of observational errors is taught today in almost exactly the form that Gauss one time gave it. But more important for future development was the fact that this theory forced researchers to make clear the conditions under which the *law of the normal distribution* is applicable. This law is often called Gauss's distribution law. Gauss however was not the first to discuss the accuracy of measurements. The founder of experimental astronomy, Tycho Brahe (1546–1601), had clearly understood that there were two groups of errors in measurements, the systematic and the random, and he attempted to eliminate the former by varying the experimental conditions for his most important observations.

Systematic errors depend upon external conditions and either upon instruments—a scale can be improperly graduated, for example—or upon methods, if one wishes to measure the height of a person accurately, for instance, the person must first take off his shoes and socks, since otherwise there will be an error no matter how carefully one measures. Random errors are errors that remain after the systematic errors have been eliminated as far as possible. The boundary between the two groups is variable, but there are always errors that are more or less inescapable, and the only thing one can do about them is to seek out theoretical methods of determining the boundaries within which they are likely to lie.

Astronomy assumed a dominant position among the experimental sciences during the 1600's and 1700's, principally because seamen needed it for their navigation. Thus it is natural that the theory of observational errors was first developed in astronomy.

The Method of Least Squares.
Priority Argument with Legendre.

The first important treatise in this area was published in 1806 by Legendre. It was called *New Methods for determination of a comet's orbit* and had a supplement entitled "*On the method of least squares.*" It is a method of obtaining the best possible average value for a measured magnitude, from many observations of the magnitude, when the measurements are found to be unavoidably different because of random errors. If the observed values are $a_1, a_2, \ldots, a_n$ and if x is an arbitrary value, we first write down the deviations or "errors" $x-a_1$, $x-a_2, \ldots, x-a_n$. Then we find the sum S of the squares of the deviations:

$$S = (x-a_1)^2 + (x-a_2)^2 + \ldots + (x-a_n)^2.$$

Finally we look for that value of x which makes the sum S of the squares as small as possible. From this comes the name "method of least squares." The sought-after value of x is the arithmetic average

$$x = \frac{a_1 + a_2 + \ldots + a_n}{n}.$$

We will shortly illustrate this method with examples.

Gauss's first publication on the method of least squares appeared in 1809 at the end of his *Theoria motus*. He mentioned there in passing that Legendre had presented the method in his work of 1806, but that he himself had already discovered it in 1795 (it should have been 1794, as we pointed out above). Legendre became disturbed, which is not at all surprising, and wrote a letter to Gauss in which he censured him. He asserted that Gauss had, as was "his method," reported a result that had been published earlier by another. Laplace acted as diplomatic mediator, and in a letter to him (which was not written until 1812) Gauss said that he had used the method almost daily ever since 1802 in his calculations concerning the new planets, and that he had communicated the method to several of his friends, for example to Olbers, in 1803.

There was never any public scientific dispute, since Gauss detested such affairs. History's judgment has been Solomonic. Gauss's correspondence and the papers found after his death proved that he was certainly first to make the discovery, but since Legendre was first to publish it, priority rights belong to the latter. Honor belongs to both of them, since they reached the result independent of each other. Concerning Gauss this is a decision that recurs several times in cases with other opponents, such as in the case of elliptic functions and the case of non-Euclidean geometry, where for the most part he made no public claim to any priority rights. He was certainly aware that his method of action would lead to complications, but he put his principle of the finished work of art before priority.

We shall now give two simple examples of the use of the method of least squares. We measure a distance, and for the sake of certainty we measure it four times. Suppose we have made no systematic errors and that we obtained the results 99, 102, 100, and 103 meters by measurements that we look upon as being carried out with equal care. Not all of these results can be correct. Naturally we want the deviation from the *true* value to be as small as possible. The "true" value, of course, belongs to Plato's world of ideas, while in reality we shall have to be satisfied with some method that, from the mathematical point of view is easy to handle and determines what we can consider the "best" or "most probable" value. Long before the theory of error ever saw the light of day, common sense had chosen as the most probable value the average M,

$$M = \frac{99 + 102 + 100 + 103}{4} = 101.$$

The method of least squares is the mathematical expression of this common sense. Let us denote an arbitrary value by x and write down the deviations $x - 99$, $x - 102$, $x - 100$, and $x - 103$. If we form the sum of the squares of these deviations, that is,

$$S = (x - 99)^2 + (x - 102)^2 + (x - 100)^2 + (x - 103)^2,$$

then to use the method of least squares we must find the value of x which makes the sum as small as possible. If we multiply out the squares in the expression for S we can then write it in the simplified form

$$S = 4(x - 101)^2 + 10.$$

Here we see that the least value for the sum of the squares is $S = 10$, which is assumed when $x = 101$, that is to say for the average value M, since the right side is always greater than 10 for other values of x.

As a second example we choose two quantities x and y, where one is dependent upon the other through a linear relationship, an equation of the first degree, which can be interpreted geometrically as a straight line. A simple case is the expansion of a railway track when the temperature rises. Here we choose purely fictitious numbers in order to have simple calculations. Suppose the linear relation we seek is

$$y = A + Bx,$$

where A and B are numbers to be determined, and that we have six pairs of values of x and y, namely

$x_1 = 2$	$x_2 = 4$	$x_3 = 6$	$x_4 = 8$	$x_5 = 10$	$x_6 = 12$
$y_1 = 2$	$y_2 = 4$	$y_3 = 4$	$y_4 = 5$	$y_5 = 5$	$y_6 = 6.$

We denote these pairs of numbers by (x_n, y_n), $n = 1, 2, \ldots, 6$, and cross in the corresponding points in a coordinate system (Fig. 13). If all of these points lie on a nonvertical straight line, then all of the coordinates (x_n, y_n) will satisfy a relation of the form $A + Bx - y = 0$. But, because of the random errors in observations, the (x_n, y_n)'s generally

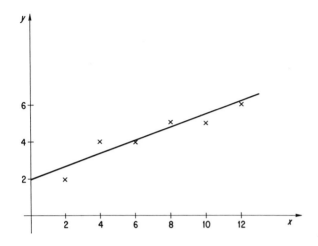

Figure 13
Determining a
linear relation (a
straight line) by
the method of
least squares.

do *not* lie on a straight line. Deviations v_n arise, which are determined by the equation

$$A + Bx_n - y_n = v_n, n = 1, 2, \ldots, 6.$$

According to the method of least squares we take for A and B those values for which the sum of the squares of these deviations is minimized; in other words, we find the minimum of the expression

$$S = (A + 2B - 2)^2 + (A + 4B - 4)^2 + \ldots + (A + 12B - 6)^2$$
$$= v_1^2 + v_2^2 + \ldots + v_6^2.$$

This time we have *two* variables A and B, and the determination of the values of A and B that make S as small as possible is more complicated in this case than it was when we had only one variable x. For problems of this type Gauss used partial derivatives, and in this special case we obtain the Gaussian normal equations

$$3A + 21B - 13 = 0$$
$$31A + 182B - 103 = 0$$

for the determination of A and B. This particular system of equations has the solution $A = 29/15$ and $B = 12/35$. Thus the relation we seek between x and y is

$$y = \frac{25}{35}x + \frac{29}{15}.$$

Geometrically this corresponds to the straight line drawn in Fig. 13. It has the property that the sum of the squares S of the deviations for this straight line is smaller than the corresponding sum for any other straight line.

If one makes many observations, where in general each one contains many digits, the calculations become difficult. Gauss introduced a special notation for them and a practical scheme of computation.

We ought to add perhaps that if the accuracy of a rather small number of observations is not especially great, there is little use in applying the method of least squares. It would be a case of shooting mosquitoes with a cannon. Instead, one chooses an appropriate straight line, in our case for example the line $y = (1/3)x + 2$, which passes through the points (6,4) and (12,6). Sighting by eye along the line we get an idea of how close to the goal we have come.

When Gauss tried to determine the orbits of planets or comets, the curve he sought was not a straight line as in Fig. 13 but an ellipse or a parabola. The problem immediately became much more difficult but he continued to use his method of least squares to determine, for example, that ellipse for which the sum of the squares of the deviations from certain observed points is a minimum (Fig. 14).

In 1823 Gauss published his great work *Theoria combinationis observationum erroribus minimus obnoxiae*—A theory for the combination of observations, which is connected with least possible error. It is a systematic and generalized presentation of his earlier theory of observational errors. Here he develops the method of least squares with mathematical rigor as, in general, the most suitable way of combining observations, independent of any hypothetical law concerning the probability of error.

The Law of the Normal Distribution. The Error Curve.

The famous Gaussian distribution law, with various applications, was

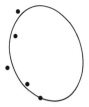

Figure 14
Determining a planetoid's orbit by the method of least squares. (The figure is not based on actual observations.)

published in the third section of *Theoria motus* (in connection with the method of least squares) and in some later papers, as for example in *Bestimmung der Genauigkeit der Beobachtungen*, Determination of the accuracy of observations, which Gauss published in 1816.

When Gauss made his derivation of the error curve he probably started from his own empirical results in measuring. We can surmise that he drew a diagram similar to that in Fig. 17 and then searched for the equation of the curve which best corresponds to the distribution of measurement errors around the arithmetical average that he had observed.

He made a number of general assumptions about the observations and the observational errors, and supplemented them with a purely mathematical assumption. Then in a rather simple way he was able to derive the equation of a curve that corresponded well with his empirical results. It was the Gaussian curve

$$\phi(x) = \frac{h}{\sqrt{\pi}} e^{-h^2 x^2},$$

where the constant h is the precision coefficient, and $e = 2.71828\ldots$ is the base of the natural logarithms.

The graph of $\phi(x)$ looks like a bell and it is sometimes called a bell-shaped curve. If the precision coefficient is large, then the curve is steep, and the observations lie closely gathered about the arithmetical average. But if it is small, the curve is flattened—the distribution is then more widespread (Fig. 15).

The probability that an error lies in a certain interval (a,b) is expressed mathematically by the integral

$$\int_a^b \phi(x)\,dx,$$

which has its geometric counterpart in the shaded area between points a and b on the x axis in Fig. 16. If we wish, we can also determine the probability of an error lying between the numbers a and b by graphically calculating the shaded area above the interval (a,b). If the curve is drawn on graph paper we can simply count the number of squares.

Now suppose that the largest possible boundaries for error are $-A$ and B, that is suppose that all of the errors x lie in the interval $-A \leq x \leq B$, where A and B are fixed positive numbers. Then the probability that all of the errors lie within this interval is equal to 1, which is the mathematical expression for complete certainty. This implies that

$$\int_{-A}^B \phi(x)\,dx = 1,$$

but since the numbers A and B cannot always be given exactly, for certainty's sake on takes $-\infty$ and $+\infty$ as the lower and upper bounds in the integral. We thus have

$$\int_{-\infty}^{\infty} \phi(x)\,dx = \frac{h}{\sqrt{\pi}} \int_{-\infty}^{\infty} e^{-h^2 x^2}\,dx = 1,$$

independent of the value of h. Naturally this result must agree with the mathematical computation of the integral. Laplace probably was the first to compute this classical integral, and Gauss used his value for the integral when he derived $\phi(x)$.

The area between the bell-shaped curve and the x axis is thus equal to 1. The fact that the curve approaches the x axis more and more closely as x goes to $+\infty$ or $-\infty$ is a geometric expression of the fact that the probability of very large error is very small (Fig. 16).

Gauss also carried out calculations of the probable error boundaries for a particular series of observations of the same quantity. The

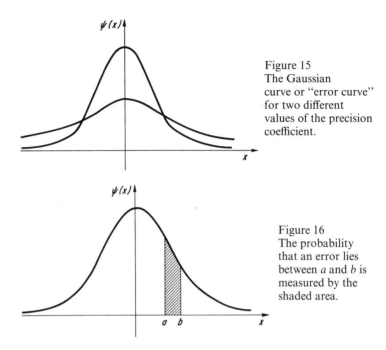

Figure 15
The Gaussian
curve or "error curve"
for two different
values of the precision
coefficient.

Figure 16
The probability
that an error lies
between a and b is
measured by the
shaded area.

most important result here is that the average error of the arithmetical average is inversely proportional to the square root of the numbers of observations; in other words, the probable accuracy of the average grows with the square root of the number of observations. Consequently it pays to make many observations, but on the other hand accuracy does not grow very rapidly through numerous repetitions.

When we wish to write down the experimental density curve we can do so in the following way: As before, we denote the different measurements by $a_1, a_2, \ldots, a_n$, and the arithmetical average of the measurements by M. The *apparent* errors according to Gauss are the deviations $M - a_1, M - a_2, \ldots, M - a_n$. Then the errors are classified with regard to sign and placed in intervals from 0 to 1, 1 to 2, and so forth, each of unit length. The percentage of errors that lie in the different intervals is then set down as the ordinates (bars on the graph).

We thus have a number of rectangles (Fig. 17) whose combined area is equal to 1. Through these rectangles we draw a curve in such a way that the shaded areas associated with each rectangle are equally large. The curve obtained in this way is the experimental error curve. Figure 17 shows an ideal *normal density curve*.

Such a curve shows up, for example, if one measures the height of all persons of a given age who are liable for military duty in a country. The heights of the different individuals are then distributed about the arithmetical average in about the same way that the errors in Fig. 17 are distributed about the vertical axis. But in reality one also runs up against skewed distributions, so that instead we may get a curve more in the style of Fig. 18, which shows how the age of men at their first first marriage were distributed during the years 1901-1910 in Sweden (according to an investigation by the statistician Sven Wicksell).

In his older days, during the years 1845-1851, Gauss wrote a detailed work on "The application of the calculus of probabilities to balance computation in widow's funds," with tables for calculating

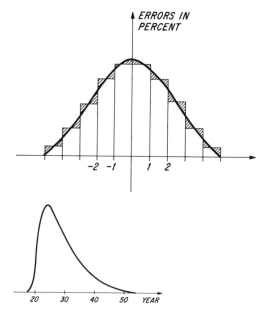

Figure 17
The Gaussian
normal curve.

Figure 18
Skewed distribution.

life annuities and other such things. He took this vast work, which became a cornerstone of modern actuarial mathematics upon himself because of the ill-planned pension conditions for the widows of professors at Göttingen. There had become too many of them and the fund was threatened with ruin. Thanks to Gauss an economic catastrophe was avoided, but the subject was not applicable to his own family, since at that time he had again been a widower for a long time.

SEVEN ⬡ GEODETIC MEASUREMENTS

The word geodesy comes from the Greek and denotes earth division. Long ago geodesy became the name of the science that deals with the form and size of the earth, with its measurement, and with the production of various maps and sea charts. The fundamental trigonometric principles of map making were already known in antiquity.

If we wish to determine the distance between two widely separated points A and B on the surface of the earth, we can do so in the following way (Fig. 19). First we very carefully measure a basic line AC, and then install angle-measuring instruments, called theodolites, at the endpoints A and C. From these points we sight point D. It should be an appropriate point in the terrain, one that can be seen in spite of the earth's curvature, such as a mountain top or a church tower. We measure the angles DAC and ACD and then solve the triangle ACD, that is to say, we calculate all of its sides and angles. Using CD as a new base line, we sight toward a point E and then solve the triangle CDE, just as we solved triangle ACD. From DE as base line we finally sight toward the goal B, and the calculation of the desired distance AB becomes the same kind of trigonometrical problem as the others.

This method for the measurement of great distances, and for map making in general, is called *triangle measurement* or *triangulation*. The places A and B are connected by straight lines that are sides of a polygon. The dashed line $ACEB$ (or ADB) is called a *polygonal* line. A net of triangles may thus be laid over the entire landscape.

Through astronomical observations one can determine the latitudes, i.e., the angular distances to the pole, for A and B. If the two places lie on the same meridian, and if one assumes that the earth is a sphere, one can use proportions to calculate the earth's circumference from the known information. For example, if the circle segment AB is

400 kilometers and the difference in latitude is 3.6 degrees, then the circumferences of the earth must be 400,000 kilometers. Eratosthenes, who did more than sieve the prime numbers, made an amazingly accurate determination of the earth's circumference by using this principle.

From the assumption that the earth is a sphere it follows that the length of one degree of meridian, that is to say an arc of a meridian that subtends an angle of one degree, must be the same everywhere on earth. But the very first measurements of this type indicated that the meridian degree is not of constant length, and the conclusion was drawn that the earth was accentuated near the poles somewhat like a lemon. Newton's law of gravitation in conjunction with the rotation of the earth, however, demanded the opposite conclusion: the earth ought to be flattened toward the poles somewhat like a tomato.

The first measurements of a meridian degree were not particularly accurate, and in addition they had been made at places that did not have sufficient differences in latitude. In order to settle the question, the French academy sent an expedition to Peru in 1735 under the leadership of Pierre Bouguer, and another to Tornedalen in Sweden in 1736 under P. L. M. de Maupertuis. The meridian degrees of Peru and Lappland were then to be compared with a meridian degree that had been measured earlier in France. If the earth is flattened at the poles, the meridian degree in Lappland at 65 degrees north latitude should be *longer* than that in France at 45 degrees latitude. The three different measurements showed definitely that the earth was shaped more like a tomato than a lemon. In technical terms the earth is nearly an ellipsoid of rotation or a spheroid, and the flattening agrees fairly well with the value that Newton had computed theoretically. His result was 1/230; the value accepted today is 1/297.

During all the eighteenth century the mapping of the earth, both on a large and a small scale, was a fundamental problem in theoretical as well as practical astronomy. Therefore it is natural that Gauss was initiated into the problems of geodesy at an early stage. In the year 1799 the Prussian Colonel von Lecoq, who was working on a map of Westphalia, turned to the twenty-two-year-old Gauss with theoretical questions about triangulation. From their exchange of letters in 1800 it is clear that Gauss already had numerous ideas at that time—such as the

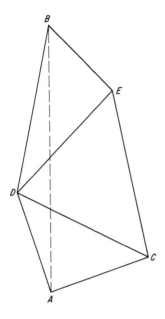

Figure 19
Triangle net for determining the
distance between two points A and B
that lie far apart.

geodetic triangle and the conformal mapping of the spheroid onto a sphere or a plane—which he was to develop much later, and which would result in the fact that to his many other masterships he could add the title "world's foremost geodesist."

During the years 1800–1805 Gauss often took part in different field work with his sextant. He was nearsighted, but this does not seem to have been any great handicap either then or later.

The Triangulation of Hannover. The Heliotrope.

The principal cause, however, of Gauss becoming a practical geodesist, was Heinrich Christian Schumacher (1780–1850), a doctor of law who had been one of Gauss's first students at Göttingen. He later became professor of astronomy in Copenhagen and director of the observatory at Altona, which then belonged to Denmark. Schumacher had been commissioned by the Danish government to make geodetic measurements throughout the country. In June of 1816 he asked Gauss for help in planning the work and at the same time expressed the wish that Gauss continue the triangulation to the south.

Gauss was quite satisfied with the proposal since it agreed with his own plans. It was his opinion that such measurements would yield "beautiful results" for the problem of the geoid, that is to say, the problem of determining the earth's actual shape and finding the ellipsoid of rotation that best approximates the shape of the earth. There was the possibility of a great project of measuring a meridian arc over approximately 15 degrees of latitude, from Jutland's northern point to the island of Elba.

At that time Göttingen belonged to the United Kingdom of Great Britain, Ireland, and Hannover under King George III. In 1820 a grant of money was provided for the triangulation of Hannover, with Gauss in charge. As assistants he chose mostly military men; the reasons are given in a letter to Schumacher in 1817: "For the most part it is best to have officers as assistants, since the peasants have the greatest respect for them, and under any conditions a certain military order is not without its uses."

Also among his helpers was his eldest son, the fifteen-year-old Joseph, who later became an artillery officer and then director of Hannover's railroads.

During 1821–1823 Gauss and his team carried out field work on a polygonal line between the observatories in Göttingen and Altona. The following two years the chain was extended west to Jever and Varel at Gauss's suggestion; by doing so they were able to connect onto the Dutch geodetic network and via it to the French network. Gauss wrote yearly reports about his work; these were collected in 1828 in his publication *Determination of the difference in latitude between the observatories in Göttingen and Altona by observations with Ramsden's zenith sector*.

During the field work Gauss proved that he was more than a desk worker. He developed significant organizational ability, great technical proficiency with instruments, a meticulous sense of order, and tenacious endurance.

The field work also resulted in a practical discovery, the heliotrope. Gauss conceived the idea when he was troubled by the reflection of the sun from a far distant windowpane. The principle of the heliotrope is the same as that used by children when they use a mirror to reflect sunlight on distant objects. The most important part is a revolving mirror, which by a simple and ingenious sighting apparatus can be maneuvered by hand in such a way that the sunlight is always reflected in a certain direction. In the measurement of triangles it is used both as a sighting goal and an optical telegraph. To an observer several miles away it looks like a clearly shining star of the first or second magnitude.

Gauss himself supposed that "under favorable atmospheric conditions there no longer exist any bounds for the sides of a triangle, except those that are imposed by the curvature of the earth." His wish for large triangles was perhaps only partly dependent upon purely practical grounds. We will return to this later in connection with non-Euclidean geometry.

With the theoretician's delight at a practical discovery, Gauss proposed a use of the heliotrope which reminds us of today's laser: "With 100 banked mirrors, each sixteen square feet in area, one would be able to send a fine heliotrope light to the moon. It is a shame that we cannot send out such an apparatus with a force of 100 men and several astronomers to give us a signal for the determination of longitude."

Altogether Gauss was out in the field during the summers for nearly ten years, in order to carry out or supervise the measurements. He also directed the continued triangulation of Hannover during the years 1828–1844. Through the use of the heliotrope and Gauss's method for measuring angles, the observations had a precision heretofore unachieved. The conversion of observational material to maps by means of new theoretical techniques was made possible through the tremendous development of geodesy as a science. In the geodetical work of

1828 mentioned earlier, there appears the first definition of the earth spheroid as a *level surface*: "What we from a geometrical point of view call the earth's surface, is nothing other than that surface, which at every point intersects the direction of the force of gravity in a right angle." With that the determination of the geoid's form became a problem in what is known as potential theory.

Gauss's theoretical work became the foundation of modern geodesy. His experimental methods were dominant for a long time, but lately they have been largely replaced by others, such as map-maping from aerial photographs. For several years now satellite observations have been used for tying together the maps of a country or a continent.

Many devotees of mathematics and mathematical physics have complained that Gauss devoted too much of his time to field work and numerical calculations; Gauss himself estimated the number of figures he handled during his geodetic measurements at more than one million. Bessel, himself a geodesist of the highest class, as early as 1823 in a letter to Gauss spoke of the "loss of time" which the latter must have suffered through his field work. Several months later Gauss replied: ". . . All of the measurements in the world do not outweigh one theorem by which the science of eternal truths is really carried forward. However you must not judge on the absolute, but on the relative value."

Here Gauss alludes to his own triangulation between Göttingen–Altona–Jever–Varel. But at that time there still remained the enormous calculational work involved in the mapping of all of Hannover; this was commented upon by Sartorius von Waltershausen, among others: "So finally there exists, as the crowning piece of this geodetical work, a coordinate list of more than 3,000 points in Hannover, in which every pair of numbers is the result of a lengthy calculation carried out by the method of least squares, and which a less competent calculator might have needed several days to carry out."

EIGHT ⬡ CURVED SURFACES

A Surface's "Inner" Geometry. The Parametric
Representation of the Equation of a Curved Surface.

Gauss's theoretical work on curved surfaces is related in two ways to his other work: on one hand to his purely practical geodetic measurements, and on the other to his investigations of the foundations of geometry.

The study of curved surfaces belongs to what is called differential geometry, which was begun by Archimedes when he, for example, determined the volume and surface area of the sphere, or when he found the area of a surface generated by a segment of a parabola. But then nearly two thousand years passed before anyone continued where Archimedes had left off when he was slain by a Roman soldier during the plundering of Syracuse in 212 B.C.

During the 1600's and 1700's progress was good, thanks to analytic geometry and differential and integral calculus. Nearly all great mathematicians made some contributions to the study of curves and curved surfaces; they studied tangents and normals, areas, volumes, lengths of curves, curvature, and so forth.

This classical differential geometry received new impetus during the decade around the turn of the century in 1800. The foremost wellspring of ideas was Gauss with his studies of curved surfaces.

He used a new principle in that he studied the curved surface's *inner properties*, that is to say, those properties that can be discovered by investigation of the surface itself, independent of the surrounding three-dimensional space. He concentrated on small regions and investigated those properties of the curved surface that could be described, for example, through very small movements along a curve on the surface.

Since then this principle has dominated the development of differential geometry and topology.

Gauss's new "inner" geometry was intimately related to his method of representing the equation of a surface in space. We shall illustrate this with two examples. We choose the unit circle first. From the chapter on the 17-gon we know that its equation can be written

$$x^2 + y^2 - 1 = 0. \tag{1}$$

But from trigonometric definitions it follows that

$$\left.\begin{array}{l} x = \cos v \\[2mm] y = \sin v \end{array}\right\}, \tag{2}$$

where v is the angle POA and P is an arbitrary point on the unit circle (Fig. 20). From the geometric viewpoint, relation (2) is equivalent to (1), and we now have the equation of the unit circle written in a new manner. On the other hand we can derive (1) from (2), since if we square and then add the two equalities in (2), we have

$$x^2 + y^2 = \cos^2 v + \sin^2 v.$$

But $\sin^2 v + \cos^2 v = 1$, and if we subtract 1 from both sides of the equation we come back to $x^2 + y^2 - 1 = 0$, which is (1).

Relation (1) has the general form

$$F(x, y) = 0,$$

which can be interpreted geometrically as a curve in the plane. If in particular we choose $F(x, y) = x^2 + y^2 - 1$ we obtain the equation of the unit circle in its usual form.

Relation (2) is of the general form

$$\left.\begin{array}{l} x = f(v) \\[2mm] y = g(v) \end{array}\right\},$$

where $f(v)$ and $g(v)$ are functions of the variable v, which in this connection is usually called a parameter. We have found a *parametric representation* of the unit circle.

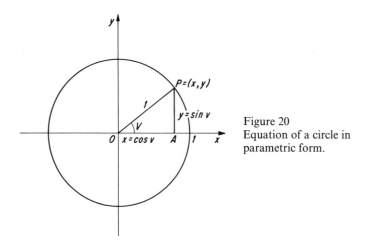

Figure 20
Equation of a circle in
parametric form.

In the differential geometry of the plane the parametric representation is superior for the study of more complicated curves, and this
superiority appears even more clearly in space.

For curved surfaces it is necessary to have three mutually perpendicular axes in the Cartesian coordinate system. We can think of these
axes as forming the edges that pass through one corner of a cube, and
can then place the bottom of the cube on the $x–y$ plane with the third
edge along the new axis, called the z axis. By the Pythagorean theorem
we used earlier we can now prove that the unit sphere, that is to say the
spherical surface with radius 1 and center at the origin of the coordinate
system, has the equation

$$x^2 + y^2 + z^2 - 1 = 0. \tag{3}$$

This is a relation of the form

$$F(x,y,z) = 0; \tag{4}$$

in a completely general way such a relation can be interpreted as a
surface in space.

But the equation for the unit sphere can also be given in parametric form. We then have

$$\left.\begin{array}{l} x = \cos u \times \cos v \\ y = \cos u \times \sin v \\ z = \sin u \end{array}\right\}, \tag{5}$$

where u is the latitude and v the longitude of an arbitrary point on the spherical surface. Relations (3) and (5) are—as were (1) and (2)—equivalent; by squaring and then adding the equalities in (5) we can obtain (3).

Relation (5) is superior when one is moving about on a sphere. We do not navigate with respect to three mutually perpendicular axes that meet at the center of the earth. Instead we use latitude and longitude, which are the "natural coordinates" on the surface of a sphere or a spheroid; they can be measured as curves on the surface itself, independent of the surrounding space. It was probably the latitude and longitude net of the earth that gave Gauss the idea of foregoing the relations of type (4) in the study of arbitrary curved surfaces, and instead using those of type (5). In their general form the latter relations can be written

$$x = f(u,v), \; y = g(u,v), \; z = h(u,v), \tag{6}$$

where u and v are two parameters, called the *curvilinear coordinates* of a point on the surface. It then becomes apparent that the fundamental quantities in the theory of curved surfaces can be expressed with the help of the parameters u and v. Relations of type (6) make it possible to follow the line of "inner" geometry consistently.

Gauss used the parametric representation of surfaces in a paper of 1822, which was the answer to a contest question posed by the Danish Academy of Science, and which was first printed in 1825: *The general solution of the problem of mapping a given surface onto another given surface in such a way that very small pieces of the image surface are similar to the corresponding pieces of the original.*

Gauss later christened a mapping with these properties *conformal*, instead of the longer expression "in den kleinsten Theilen ähnlich"—"similar in the small" which he had used at first. That a mapping is conformal implies that angles are preserved; in other words, it retains the angle between two curves from the original to the image. If, for example, the angle between two roads in a landscape is 30°, then the angle between the roads on the map will also be 30°.

Angle-preserving mappings are both very old and very useful. Mercator's projection dates from 1569 and continues to be used on sea charts. The *loxodrome* between two points is a curve that intersects the meridians along the way at a constant angle. Thus it represents a compass direction between the two points. Under Mercator's projection the loxodromes are mapped as straight lines. Thus a ship that holds a constant course follows a straight line on the sea chart.

In his work on conformal mapping Gauss first handled the general problem and then gave several examples of projections closely related to his geodetic measurements: the mappings of the plane, the cone, the sphere, and the spheroid onto a plane, and finally the mapping of a spheroid onto a sphere.

For the conformal mapping of one region onto another, the *analytic functions*, which we shall mention later, play an important role. The theory of analytic functions was conceived by Gauss but developed by others.

The Curvature of Curves in the Plane and on Curved Surfaces.

Gauss's most important work in differential geometry is one printed in 1827: *Disquisitiones generales circa superficies curvas*—General investigations of curved surfaces.

The definition introduced there of the radius of curvature of a surface at a given point on the surface plays a fundamental role in the theory of curved surfaces. To illustrate this notion we begin with the curvature of a curve drawn in the plane (Fig. 21). We start with a plane curve *APB*. It should satisfy certain mathematical conditions which, however, we will not go into here. When we are to determine the "curvature" of the arc at an arbitrary point *P* we proceed from our intuitive notion of the curvature of a circle: a circle with large radius has small curvature, and a circle with small radius has large curvature.

A circle is determined by three points. Therefore we pass a circle through point *P* and through two other arbitrary points P_1 and P_2 on the curve. The circle has its center at a point *Q*. Now we let points P_1 and P_2 move in toward *P* along the curve. Then we obtain a series of different circles with different centers. Passing to the limit, when both P_1 and P_2 coincide with *P*, the point *Q* falls at *O*, and we then have a circle with radius $OP = r$, which is tangent to the curve at *P*. It is called the *curvature circle* at point *P*, and *r* is called the *radius of curvature* at point *P*, while *O* is the *center of curvature*. The *curvature k* of the curve at point *P* is finally defined as the reciprocal value of the radius of curvature:

$$k = \frac{1}{r}.$$

If the equation of the curve is known, then one can derive a mathematical expression for *r*. Applying this method to a circle we find, as we expected, that the curvature circle of the circle is the circle itself.

A tangent to a curve is, in a certain sense, a straight line that intersects the curve in *two* coincident points. The curvature circle, in the same sense, intersects the curve at *three* coincident points. Since this is a question of a more or less intimate relationship, the curvature circle is often called the *osculating circle* (from the Latin *osculari*, kiss).

The tangent to the curve at *P* is also a tangent to the curvature circle at *P*. The line *PO*, which is a radius of the curvature circle, is at the same time a *normal* to the curve, that is to say it is perpendicular to the tangent of the curve at *P*.

It is clear from Fig. 22 that the osculating circle corresponds to our intuitive notion of curvature; in the figure we have also drawn in the tangents and normals at two points. It is also evident that the curvature can have different signs at different points on the curve. If we decide

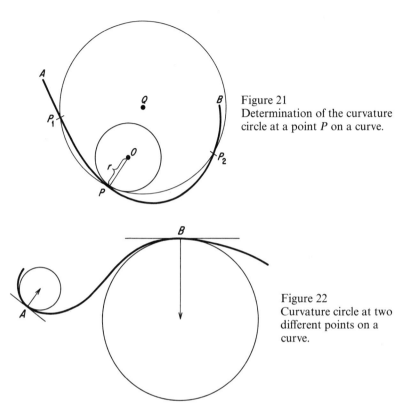

Figure 21
Determination of the curvature
circle at a point P on a curve.

Figure 22
Curvature circle at two
different points on a
curve.

that k is to be positive at point A, then k will be negative at point B, because of the fact that the two curvature circles lie on different sides of the curve. The curvature can also be zero at some points on a curved arc. If $k = 1/r$ is zero at every point, then the curve is a straight line; one may look upon a straight line as the limiting case of a circle, where the radius r "goes to infinity."

Gaussian Curvature.
When we wish to study the curvature at an arbitrary point P on a curved surface in space, we first draw the normal PN to the surface at

point P. (Fig. 23). It is perpendicular to the surface's tangent plane at point P. We then consider the different curves of intersection of the surface that are cut out on the surface when a plane that contains the normal line PN is rotated about the normal line. These normal intersections will be different plane curves that in general do not have the same curvature at P. If all of the curvatures are not equal, there will be two extreme values among them, a maximum and a minimum, the so-called *principle curvatures* $1/R_1$ and $1/R_2$, where R_1 and R_2 are the *principle radii of curvature*. The corresponding normal intersections are perpendicular to each other. In Fig. 23, curve APB is one of the normal intersections, and the oval with PC as its axis is the other. The product

$$K = \frac{1}{R_1} \times \frac{1}{R_2} = \frac{1}{R_1 R_2}$$

is the *Gaussian curvature* of the surface at P or, to make a long story short, the surface's curvature at P.

In our example R_1 and R_2 have different signs, since the corresponding centers of curvature lie on different sides of the surface; thus K is negative. If this holds for every point on the surface then we say the surface has *negative curvature*. One such surface is the hyperboloid of rotation with one mantle that we drew in Fig. 23. It resembles an hourglass and is generated by revolving the hyperbolic arc APB around the axis EF.

If R_1 and R_2 are directed to the same side at each point on the surface, then K always has positive sign and we then have a surface with *positive curvature*, for example, a sphere, where $R_1 = R_2 =$ the radius of the sphere (if the positive normal direction goes in from the surface toward the center).

If $K = 0$ everywhere, then the notion of curvature in Gauss's sense disappears. Some examples of surfaces with *zero curvature* are planes, cylinders, and cones.

The curvature K can thus be used to classify curved surfaces, and in the discussion of geometries, including the non-Euclidean, this classification plays an important role.

In a schematic formulation we can say that differential geometry

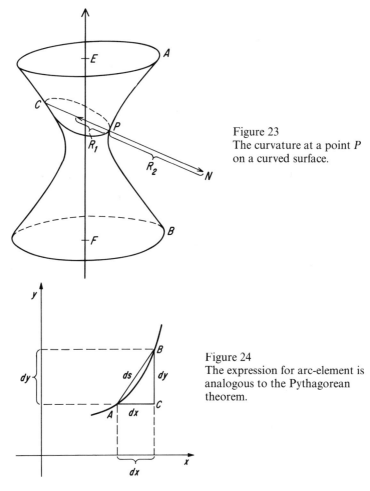

Figure 23
The curvature at a point P
on a curved surface.

Figure 24
The expression for arc-element is
analogous to the Pythagorean
theorem.

includes the study of curves and surfaces in the *immediate vicinity* of a
point, where consequently all of the distances are small. For a small segment of a curve in the plane, a so-called arc-element ds, we can write

$$(ds)^2 = (dx)^2 + (dy)^2$$

in analogy with the Pythagorean theorem (Fig. 24).

Here the expressions dx and dy denote very small changes in the x coordinate and the y coordinate, respectively, of a certain curve. Point A has coordinates (x, y), and B has coordinates $(x + dx \ y + dy)$. Leibniz called these small changes *differentials* and they have given name to both differential calculus and differential geometry. The arc element ds is also a differential, which purely geometrically denotes the hypotenuse in the right triangle ABC, and not a piece of the curve. But if the curve satisfies certain general conditions and if points A and B lie very close to one another, then one may let the chord AB replace the arc AB, and the error will not change the result appreciably. The element of arc ds will then be a sufficiently accurate measure of the rate of change in the curve's arc length s, which is caused by small changes in the x and y coordinates.

For a curve on a surface $F(x, y, z) = 0$ the square of the length of the arc-element is defined in the same way:

$$ds^2 = dx^2 + dy^2 + dz^2, \tag{7}$$

where for the sake of simplicity we have now removed the parentheses. If we represent the surface in parametric form

$$x = f(u,v), \ y = g(u,v), \ z = h(u,v),$$

ds^2 goes over to the quadratic form

$$ds^2 = E \, du^2 + 2F \, du \, dv + G \, dv^2, \tag{8}$$

where the coefficients E, F, and G are functions of the parameters u and v. Exchanging the simple relation (7) for the more complicated relation (8) does not seem to produce any advantage. But it is (8), the so-called first fundamental form, which is the gateway to the inner geometry of of a surface.

Invariants.

Before we go farther we want to mention a concept that is important in all of mathematics, that of *invariants*. Quite generally, invariants are mathematical expressions that are not changed under certain transformations of the variables that make up the expression. We may say that invariants are descriptions of the natural laws of mathematics.

A simple example of an invariant is the distance between two points in a plane. It is not changed when we rotate the coordinate axes or when we move them parallel to their original positions. The equation of the line segment joining the two points is changed, as are coordinates of the end points of the line segment, but the distance itself remains unchanged.

Another example is given by a notion that we mentioned earlier, the conformal mapping of one surface onto another; for example, Mercator's projection, where the angles are the invariants.

A mapping of one surface onto another that leaves the arc-elements in relation (8) invariants is said to be *isometric*. In this case all of the arcs that correspond to one another under the mapping have the same length.

From a mechanical point of view an isometric mapping entails a change in form of the surface in which one goes from the original to the image in such a way that the surface *bends without stretching*—it may be neither drawn together nor stretched out in any direction. If a rectangular paper is bent into a right circular cylinder, an isometric mapping has been carried out. All of the quantities that remained unchanged under an isometric mapping are called *isometric invariants*.

In his paper on curved surfaces of 1827 Gauss proved that the curvature K is an isometric invariant. This theorem has played a central role in both differential geometry and the general theory of relativity. If, for example, the hourglass in Fig. 23 had been constructed of a flexible and thin (strictly speaking, infinitely thin) piece of glass, then we could bend the glass without stretching it. We would then have the same values of K at any given point on the original surface and the corresponding point—the image point—on the deformed surface. If the first principal radius of curvature R_1 is increased by the bending then the second one, R_2, is decreased in such a way that the product $R_1 R_2$ remains constant.

Gauss himself called this remarkable result the "*theorema egregium*"—an extraordinary theorem. Perhaps he was thinking of it in 1825 when he wrote in a letter to the astronomer Peter Hansen: "These investigations deeply affect many other things; I would go so far as to say they are involved in the metaphysics of the geometry of space."

Another important invariant under isometric mappings is the length of what are called *geodesic lines*. One can draw infinitely many curves between two points A and B on a surface. That curve for which the length of the arc AB is a minimum, is called the geodesic line. Thus it gives the shortest path between two points on a surface. If the surface is a plane then the geodesic line will be a straight line. This is one of the axioms in Euclid's geometry.

For the geometry of curved surfaces the geodesic lines play the same role as the straight lines do in plane geometry. The geodesic line between points A and B on a surface—like the straight lines in the plane—is *uniquely* determined by the two points; at least this is true if the two points lie sufficiently close together. (Two diametrically opposite points on a sphere, for example, have no uniquely determined geodesic line.)

If a surface is bent without stretching, then the geodesic lines in the original are carried over to geodesic lines in the image. This is true because according to the hypothesis all curves retain their length, and thus the shortest becomes the shortest. Consequently the geodesic lines are isometric invariants.

In the case where the surface is a sphere, the geodesic line between two points will be an arc of a great circle (a great circle is the intersection of the sphere with a plane that passes through the center of the sphere, such as for example, the equator and all of the meridians). Flights over the polar regions are relatively new practical applications of geodesic lines.

Geodesic Triangles.

A *geodesic triangle* is constructed from the geodesic lines between three

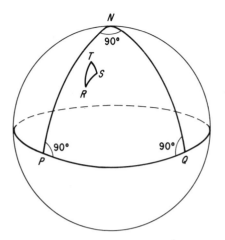

Figure 25
Two triangles drawn on a sphere. The sides are segments of great circles. The sum of the angles in a spherical triangle is always *greater* than 180°.

points on a surface. An angle of a geodesic triangle is the angle between the corresponding arcs. The angle between two curves is the angle between the tangents to the curves at the point of intersection. In the plane, the geodesic triangle is the "usual" triangle; the sum of its angles is 180°. But on a spherical surface it is made up of arcs of three great circles.

Figure 25 shows a geodesic triangle *NPQ* that is constructed from the equator and two meridians with a longitudinal difference of 90°. The sum of the angles of this triangle is 270°. If we rotate the meridian *NP* so that the difference in longitude between *NP* and *NQ* is only 1°, then we have a smaller triangle and a smaller sum for the angles, namely 181°. But no matter how small a triangle we draw on a sphere (for example *RST* in Fig. 25) the sum of the angles will always be *larger* than 180°.

One of the most famous results in Gauss's theory of surfaces concerns the sum of the angles in a geodesic triangle. On a given surface we choose three points *A*, *B*, and *C* which lie sufficiently close to one another so that the geodesic triangle is uniquely determined. We may also let *A*, *B*, and *C* denote the measurements of the respective angles of the triangle. We now measure an angle, not in degrees as was done earlier, but in radians, that is to say, as the length of the arc that the angle subtends on the circle of radius 1. Since this circle's radius is 2π, we find that 360° corresponds to 2π radians, 180° corresponds to π radians, and so forth.

In general, the sum of the angles in a geodesic triangle ABC deviates from the "Euclidean" value of π radians (or 180°). The angle deviation, denoted by V, is defined by

$$V = A + B + C - \pi. \tag{9}$$

If V is positive we have a surplus, an angle excess, and if V is negative we have a loss, an angle deficiency.

In his paper on curved surfaces of 1827 Gauss derived a formula for V, in which the curvature K plays a decisive role. If the surface has constant curvature, this formula is quite simply

$$V = K \times T, \tag{10}$$

where T denotes the area of the triangle. By combining (9) and (10) we then have

$$A + B + C - \pi = K \times T. \tag{11}$$

In the general case, when K is not constant, the V in equation (10) is expressed as an integral which, according to Gauss's definition, is the total curvature of the area ABC. The result can thus be formulated as follows: *The angle deviation is equal to the geodesic triangle's total curvature.*

After having proved this, Gauss, who for want of equals became his own critic more than once, said: "This theorem is seen to be one of the most elegant in the theory of curved surfaces."

Generalizations.

In the next chapter we shall discuss the applications of relation (11) to non-Euclidean geometry. To conclude this chapter we shall return to the meaning of a surface's parametric representation

$$x = f(u,v), \; y = g(u,v), \; z = h(u,v).$$

This implies that only *two* coordinates u and v are needed to give the position of a point on a surface whose form is given in the sense that we know the functions f, g, and h. In the same way only *one* coordinate v is necessary to give a point's position on a plane curve whose form is given in the sense that we know the curve's equation in parametric form,

$$x = f(v), y = g(v).$$

Because only *one* curve coordinate is needed to give the position of an arbitrary point on a curve in the plane, the plane curve is called a *one-dimensional point set*. Analogously, a curved surface is called a *two-dimensional point set*. One can continue this same point of view of abstract sets in 3 dimensions, 4 dimensions, and so forth, and obtain *n-dimensional sets* for each positive integer n.

This extension to several dimensions was carried out by, among others, the great German mathematician Bernhard Riemann (1826–1866). The generalization of Gauss's theorem on the invariance of curvature led him to a classification of general geometric spaces by use of differential expressions analogous to relation (8). From this, in due time, followed the universal invariant theory which is also called the general theory of relativity.

About one hundred years after the publication of "General investigations of curved surfaces" the work was judged by an equal of Gauss. Einstein says: "The importance of Gauss for the development of modern physical theory and especially for the mathematical fundament of the theory of relativity is overwhelming indeed; . . . If he had not created his geometry of surfaces, which served Riemann as a basis, it is scarcely conceivable that anyone else would have discovered it."

NINE NON-EUCLIDEAN GEOMETRY

Euclid's Geometry. The Parallel Axiom.

Euclid gathered together the geometrical knowledge of his time in his *Elements*, which completely dominated the schools' instruction in geometry, almost to the present day, and in so doing set a hard-to-beat bestseller record for textbooks.

Euclid starts from certain fundamental definitions, postulates, and axioms; then the various theorems follow from these, one after another. Euclid's goal was a purely logical method of proof, without the aid of geometric figures. He succeeded to such an extent that Euclidean geometry is one of the most admirable systems of thought ever constructed.

The method of presentation in the *Elements* became a model of logical presentation. When Gauss or Newton speak of "*rigor antiquus*"—the rigor of antiquity—in method of proof, they certainly have Euclid in mind more than Archimedes, for example. It is another matter when they put Archimedes in a higher class as a creative mathematician, presumably in the same class as themselves. The only criticism that Gauss ever directed toward Archimedes was that he could not understand why Archimedes had not discovered the decimal system.

In spite of its abstract character we should not forget that Euclid's geometry has its roots in reality. It is the geometry that arises from practical experience. We can follow it clear back to the Pharaohs' surveyors, the so-called *harpedonaptai*—rope-stretchers—who measured both arable land and the pyramids.

For 2,000 years it was the traditional view that Euclid's system was the only one that was logically and empirically correct, the only possible foundation of geometry. But there is one vulnerable point: Euclid defined parallel lines as two lines in the same plane that never meet, no matter how far one extends them in both directions. In connection with this we have the famous fifth and last axiom, the "parallel axiom." It is usually formulated more briefly today, but the meaning is the same: *Through a point not on a given straight line there exists only* one *straight line that is parallel to the given line.*

What is the empirical basis of this axiom? It did not exist during antiquity nor does it exist today. In practice we cannot extend a line to unlimited length. We can extend it 10 meters or 100 kilometers, but we cannot extend it to infinity. Even a light ray from a star is not an infinitely long line.

Perhaps Euclid had a feeling that the parallel axiom held a special position, for he placed it last and delayed using it. And his intuition did not deceive him. He placed its assertion among the axioms, and 2,000 years later it became apparent that it was correct to do so.

Certainly the parallel axiom is supported by "common sense." That is true of all of Euclid's geometry, which is still completely sufficient for the needs of practical life. Common sense, however, according to Einstein, is "as a matter of fact, nothing more than layers of preconceived notions stored in our memories and emotions, for the most part before age eighteen."

It is the task of the genius to carve into these layers, but people were wide of the mark for a long time in the matter of the parallel axiom. The reason was that the problem was attacked from the wrong point of view. In every logical system there is a rule of economy concerning the reasoning: one should not introduce more laws than are necessary; for example, one should not set up an axiom that later proves to be a theorem which can be derived from the other axioms.

For a long time it was believed that the parallel axiom was such a superfluous axiom. Over the centuries an enormous amount of ingenuity has been devoted to the proof of the parallel axiom as a theorem, just like the other theorems of the *Elements*. But it was fruitless effort. Instead one can now prove in Euclidean geometry that the parallel axiom cannot be *derived* as a theorem from the other axioms of Euclid's geometry. On the other hand, the parallel axiom can be *exchanged* for certain other suppositions that are theorems in Euclid's presentation, and one then obtains his parallel axiom as a theorem in a geometry that is still Euclidean.

Here we shall only mention one of those axioms that are equivalent to the parallel axiom: *The sum of the angles in a plane triangle is 180°*. This is a well-known theorem in Euclid's presentation. If we choose it as an axiom in place of the parallel axiom, then the parallel axiom becomes one theorem among all the others. But other than that the exchange has no consequence, and the system has the same degree of freedom from contradiction as before. The discovery of the axioms equivalent to the parallel axiom are the positive results of the unsuccessful attempts to derive the parallel axiom as a consequence of the other axioms.

As a matter of fact the parallel axiom is Euclidean geometry's mark of distinction. This became clear gradually when the problem was turned around by asking this question: If we discard the parallel axiom or an axiom equivalent to it and replace it with another—for example the axiom that the sum of the angles of a triangle is *less* than 180°—then what does the new geometry look like?

It was a dizzying, almost indecent question, which contended with common sense and with Newton's and Kant's enormous authority, all of which vouched for the Euclidean nature of space. The force of general opinion was added to the mathematical and scientifical difficulties.

Gauss's Results in non-Euclidean Geometry.

Such was the situation when the students Gauss and Wolfgang Bolyai discussed the foundations of geometry during the year preceding the turn of the century in 1800 (cf. p. 41). Of course they were not alone in thinking about the matter: time had ripened for this 2,000-year-old problem. But we must concentrate on Gauss.

His successive penetrations into non-Euclidean geometry can be found in his letters, in his friends' statements, and in the papers left behind after his death. We shall illustrate this with quotations from letters. The first one explains a short notice in his diary in September, 1799: "*In principiis Geometricae egregios progressus fecimus*"—We have made excellent progress in the question of the foundations of geometry.

Bolyai, who had returned to Hungary, thought that he had proved the parallel axiom from the other axioms. Gauss did not wish to hurt his friend, and in December of 1799 he wrote to him: "I myself have come far in that area (although my other quite heterogeneous activities scarcely leave me any time for the matter); but the way in which I have proceeded does not lead to the desired goal, the goal that you declare you have reached, but instead to a doubt of the validity of geometry. I have certainly achieved results which most people would look upon as proof, but which in my eyes prove almost *Nothing*; if, for example, one can prove that there exists a right triangle whose area is greater than any given number, then I am able to establish the entire system of geometry with complete rigor. Most people would certainly set forth this theorem as an axiom; I do not do so; though certainly it may be possible that, no matter how far apart one chooses the vertices of a triangle, the triangle's area still stays within a finite bound. I am in possession of several theorems of this sort, but none of them satisfy me."

The possibility to which Gauss alludes is realized in non-Euclidean geometry, where a triangle's area is always finite. The hypothesis that there are triangles with arbitrarily large area is, as a matter of fact, equivalent to Euclid's parallel axiom. The letter shows how deep the twenty-two-year-old Gauss had already penetrated into the fundamental problem of geometry.

Also pertinent to the widespread discussion of the parallel axiom around 1800 was the fact that if the axiom proved false, then the existence of an absolute unit of length would follow as a consequence. This had been perceived by Johann Lambert and Adrien Legendre, among others, and since one could not find such a unit, Legendre considered this in 1794 to be a strong reason to believe in the validity of the parallel axiom.

Gauss had a different opinion, which he first expressed much later, in 1816, in a letter to that astronomer Gerling. On the basis of Legendre's investigations he says that the existence of an absolute unit of length certainly appears paradoxical, but that for his part he cannot find that it implies a contradiction. He then continued: "It would even be desirable [!] that Euclid's geometry should be false, since we would then be a priori in possession of a universally valid unit of length." He then adds that such a unit of length might, for example, be the side of an equilateral triangle whose angles are 59° 59′ 59.999″.

It was certainly with intent that Gauss chose a triangle, whose sum of angles is so extremely close to 180°. Such small deviations could scarcely be noticed in surveying measurements, but that does not exclude the possibility that the space of the universe is patterned after a non-Euclidean geometry.

Two very talented jurists, F. C. Schweikart and his nephew F. A. Taurinus, made substantial contributions toward solving the mystery of the parallel lines. In 1807 Schweikart wrote a paper, in which he expressly pointed out the possibility that deviations from Euclid's geometry might only be observable in great cosmic distances; therefore he christened the non-Euclidean geometry "astral geometry." When Gauss heard this term discussed much later on, he liked it and at times used it himself.

Taurinus went much farther than his uncle and derived, among other things, the "absolute length" of non-Euclidean geometry and its trigonometric formulas; he obtained it by giving the radius of the sphere an imaginary value in the usual formulas of spherical trigonometry (as one still does). In 1824 he sent Gauss an attempted proof of the parallel axiom. In November of the same year Gauss wrote him a letter in which he set forth how far he had reached: "The hypothesis that the sum of the angles of a triangle is less than 180° leads to a strange geometry, completely different from Euclid's system, that is completely coherent, and which I have set forth for myself in a rather satisfactory manner, so that in this geometry I can solve every problem under the assumption that a certain constant is given, which cannot be determined a priori. The greater this constant, the closer one is to Euclidean geometry, and when the constant becomes infinitely large the two geometries coincide. If the non-Euclidean geometry should be true, and if the constant is not altogether too large with respect to those quantities that can be obtained by our measurements of the earth or the heavens, then this constant could be determined a posteriori."

The constant, or the absolute unit of length, which perhaps could be determined in an empirical way if non-Euclidean geometry was valid, stands in very close relation to the Gaussian curvature K, which we discussed in the preceding chapter. In fact its value is

$$\frac{1}{\sqrt{|K|}}$$

That the constant goes to infinity implies that the curvature goes to zero. On a plane surface, $K = 0$. ($|K|$ denotes the numerical value or absolute value of K; for example, $|-4| = 4$ and $|5| = 5$.)

Gauss did not explititicly mention this fundamental relation, but it is difficult to imagine that he was not fully aware of it. Probably he did not want to insert the disputed non-Euclidean geometry into his paper on curved surfaces of 1827, which is only concerned with Euclidean geometry.

In 1816 Gauss had reviewed two articles on the parallel axiom in a scientific journal. He was critical and spoke among other things of "vain attempts to conceal the gulf, which one cannot conceal, with an unsound web of sham proof." Because of these reviews he was, as he wrote to Schumacher much later, subjected to vulgar attack, which disturbed him very much. In 1829 he took up the same matter in a letter to Bessel in reply to his friend's urging that he publish his results in non-

Euclidean geometry: "Perhaps it will not happen during my lifetime, since I fear the Bœotian's cries if I were to express my opinion clearly."

Johann Bolyai. Nikolai Lobachevski.

There are two other mathematicians who are usually put in first place as the pioneers of non-Euclidean geometry: the Hungarian Johann Bolyai (1802–1860), Wolfgang's son, and the Russian Nikolai Lobachevski (1793–1856). Here we shall only touch upon their relation to Gauss.

Johann Bolyai, who became an artillery officer, was introduced at an early age to the fundamental problem of geometry by his father. Papa Wolfgang followed tradition, but the son set out upon revolutionary paths. As early as 1820, while he was a student at the Engineering Academy in Vienna, he had informed his father of his plans. Wolfgang became alarmed, and seriously warned his son Johann: "Do not waste even one hour's time on that problem. It does not lead to any result; instead it will come to poison all of your life. For hundreds of years hundreds of the world's foremost geometers have cogitated without having succeeded in proving the parallel axiom, as long as they refrained from taking some new axiom as help. I believe that I myself have investigated all conceivable ideas in this connection."

His father's advice was founded on bitter experience, but Johann did not follow it. Instead, in November of 1823 he informed his father that he had succeeded in constructing a new system which differed from Euclidean geometry; he summed up his results in a magnificent formulation: "From nothing I have created a new world."

It was not an empty assertion made in the recklessness of youth, even though his theory still was not completely finished. This beautiful new world was *hyperbolic geometry*, which results when one replaces Euclid's parallel axiom with the following axiom: *Through a point not on a given straight line, one can extend at least* two *straight lines that do not intersect the given straight line.* (Here, as before, it is a matter of geometry in the plane.) This axiom implies that *the sum of the angles in a plane triangle is less than* 180°, and that there exists an absolute unit of length.

The name hyperbolic geometry was introduced much later by the outstanding German mathematician Felix Klein (1849–1925). It comes from the geometric analogy with the hyperbola, which has "two infinitely distant points."

Johann Bolyai did not publish his "absolute geometry" until 1831, and even then it appeared merely as an appendix to one of Wolfgang's papers. Both were sent to Gauss for his appraisal. In February of 1832 Gauss wrote to Gerling about Johann's work and concludes: "I believe that this young geometer v. Bolyai is a genius of the first rank."

In March of 1832 Gauss wrote a long letter to Wolfgang Bolyai: "Now a few words concerning your son's work. If I begin by saying that I cannot judge it, you will be surprised. But I can do nothing else; praise would signify self-praise, since all of the paper's contents, and the way your son has attacked the matter, coincide almost completely with my own reflections which I partly carried out thirty to thirty-five years ago. In fact I am extremely surprised by it. My intention was to leave my own work, of which at the present time only a small part is written down, unpublished during all of my lifetime. . . . On the other hand I had intended to write it all down little by little, so that it at least would not disappear with me. I am thus quite surprised that I can spare myself these efforts, and it makes me very happy that it is the son of one of my old friends who has come ahead of me in such a remarkable manner."

Continuing, Gauss goes into some details of Johann's paper and then offers as a "specimen" his own derivation of the area of a hyperbolic triangle (which he had certainly made much earlier). He also gave Johann a hyperbolic nut to crack: Determine the volume of a tetrahedron in the new geometry. One should expect, said Gauss, that this volume can be calculated with an expression as simple as that for triangles, but such does not seem to be the case. (They both solved the problem.)

Gauss also sent along his greetings to Johann, whom he had not met: "In any case I ask you to greet him heartily from me, with the assurance of my special esteem."

Wolfgang Bolyai was quite satisfied with the letter. But the feeling was not shared at all by his son. He looked forward with soaring expectations to Gauss's judgement of his results, which he had worked out independent of anyone, even of Gauss and his own father. He was hurt when Gauss expressed the opinion that he had found the theory earlier —which without question was the case. He was not satisfied with Gauss's obliging assurance of his "special esteem" either, and he became embittered while waiting in vain for some public recognition on Gauss's part. For such recognition failed to appear. Gauss never published a review of Johann Bolyai's pioneer work, nor did he ever publish any of his own contributions to the same area. These two facts are certainly not independent of each other.

That Gauss did not publish his work in hyperbolic geometry can scarcely, because of their advanced stage, depend upon his principle of the finished work of art. Nor, on the basis of Gauss's enormous prestige, should the "Bœotians' cries" have become especially noisy, since asses do not scream against authority. The deciding reason for Gauss's silence in this matter, other than lack of time, is probably that he did not want to be involved in a new priority argument, least of all with the son of one of his oldest and best friends.

And now the other side bobs up. Could Gauss have done his friend from youth any better service than to have publicly announced his son's epoch-making achievement? The poor Bœotians' cries would not have reached any great volume then, either. But perhaps Gauss did not want to review Johann's work without at the same time showing in some way that he had achieved approximately the same results earlier even though he had not published them. One can do such a thing in private remarks, where one can support oneself with earlier letters, but it cannot be done in a public review.

From an objective viewpoint one can hardly criticize Gauss for his behavior, since he privately gave Johann Bolyai his unreserved recognition in the letter quoted. But from a subjective viewpoint Gauss would have been greater in character if in addition he had given recognition in a public review. In any case his allergy to official debate was rather unwarranted at this time. A few lines by Gauss in a scientific journal would have made Johann Bolyai famous.

But it was not to be. Johann, who resigned as a captain-lieutenant in 1833, returned from his beautiful new world to a reality that was, for him, torn asunder. During the remainder of his life he lived in Maros-Vásárhely or on the family estate Domáld. His bitterness nearly turned into paranoia when, after Lobachevski became famous, he suspected that Gauss lay behind an intrigue to deprive him of the honor of being recognized as the discoverer of non-Euclidean geometry. When his glory finally arrived, he had long been dead.

As early as 1815, as professor at the university in Kasan, Lobachevski had lectured on the foundations of geometry, with Legendre's investigations as his point of departure. During the 1820's he became convinced of the impossibility of proving the parallel axiom. In common with Gauss and Johann Bolyai, but independent of them, he began to construct hyperbolic geometry, which he at first called "imaginary geometry" and later "pangeometry."

His first paper on non-Euclidean geometry, On the Principles of Geometry, was submitted to the academy in Kasan in 1826, but it was not printed until 1829–1830. It became the first *published* paper on non-Euclidean geometry. It was followed by several others; his last work, *Pangeometry*, appeared in 1855. A linguistic iron curtain is thought to have contributed to Lobachevski's isolation from other European researchers, and vice versa. It was penetrated for the first time in 1837, when Lobachevski wrote an exposition of his theory in German in "Crelle's Journal."

Gauss seems not to have known of Lobachevski's research before about 1840. Thanks to his knwoledge of Russian he was able to read Lobachevski in the original. In a letter to Schumacher he asserted that they were the same ideas that he himself had at one time developed: "You know that during the past fifty-four years, ever since 1792, I have had the same conviction (there are several additions in connection with it, which I will not mention here). Thus I have not found anything new in the material sense in Lobachevski's work, but he develops his thoughts along a different path than I had set out upon, in a masterly fashion, in real geometrical spirit."

As we see, Gauss also gave his full recognition to Lobachevski. But this time he did not stop with a private expression by letter; instead, he also saw to it that in 1842 Lobachevski became a corresponding member of the Royal Scientific Society of Göttingen.

The honor of discovering hyperbolic geometry is thus shared by Gauss, Johann Bolyai, and Lobachevski. The extent of the share that is given to each is mostly a problem for international prestige hunters. One solution, in harmony with the character of international science, is to make the shares equally large.

Non-Euclidean Geometries and their Illustrations on Surfaces.

By non-Euclidean geometry we have so far only meant hyperbolic geometry, which after more than two thousand years was realized by three different persons in a couple of decades, each of them unaware of the other's research. This in itself is not so remarkable. Corresponding things happen quite often in the history of ideas, and they have of course given rise to the expression "the time is ripe." But what really is remarkable is that Gauss, Bolyai, and Lobachevski, all three, chose hyperbolic geometry. It is as if only this possibility, in some meta-physical fashion, should have been crystallized in the flood of time, there to be perceived by especially sensitive persons.

For there certainly does exist another possibility for changing the parallel axiom. Instead of assuming that through a point not on a given straight line there are at least two (in fact infinitely many) straight lines, one may assume there are none at all. Then in place of the parallel axiom we have the following axiom: *Through a point not on a given straight line one cannot extend a straight line that is parallel to the given line.*

This axiom results in still another non-Euclidean geometry, which turns out to be just as free of logical contradiction as the Euclidean. It came forth in the general reflections of Riemann, based upon differential geometry, in his famous trial lecture for his appointment as Docent, "On the hypotheses that are the basis of geometry" (cf. p. 105). The subject was chosen by the seventy-seven-year-old Gauss, who was curious to know how the student, fifty years younger, would handle such a difficult problem. After the lecture Gauss enjoyed one of his very rare outbursts of enthusiasm.

Because of its analogy with the ellipse, which has no "infinitely distant points," this geometry was called *elliptic* in Klein's nomenclature. Since the parabola has *one* "infinitely distant point," Euclidean geometry is also called *parabolic* geometry. Now we can set up the following scheme:

Founder	Geometry	Number of lines extending through a point not on a line, without meeting the line	Sum of the angles of a triangle
Euclid, 300 B.C.	parabolic	1	180°
Gauss, Bolyai, Lobachevski, 1800–1835	hyperbolic	At least 2	Less than 180°
Riemann, 1854	elliptic	0	Greater than 180°

It should be pointed out that in addition to these we can produce as many geometries as we wish, so long as we see to it that the freely proposed axioms do not lead to contradictions.

One can realize the hyperbolic and elliptic geometries through different models that belong to the usual Euclidean geometry. The object of such models is not primarily to illustrate the non-Euclidean geometries but rather to check their freedom of contradiction. For the models can be studied with Euclidean geometry, which in turn can be studied by the usual analytic geometry, and through this the decisive question of the non-Euclidean geometry's freedom from contradiction is carried over to the question of arithmetic's freedom from contradiction.

Here we shall only deal with illustrations using models taken from the theory of curved surfaces. From the preceding chapter we know that the straight lines in the plane correspond to the geodesic lines on a curved surface and that a triangle in the plane corresponds to a geodesic triangle of the surface.

In Euclidean geometry we assume that a figure may be moved wherever we wish in the plane without being changed. This freedom of movement is necessary, since by definition, congruent figures are those figures that can be placed on top of one another in such a way that they exactly cover each other. The same degree of freedom of movement must exist on the curved surfaces, which are the counterparts of the Euclidean plane in non-Euclidean geometry; *only* the parallel axiom should distinguish the two geometries.

For two geodesic triangles ABC and $A_1B_1C_1$ on a curved surface (Fig. 26), one of the two triangles should exactly cover the other after a displacement on the surface itself. We can think of the triangles as being cut out of a flexible sheet. The curvature K is certainly an isometric invariant. In order to be able to place the triangles on one another the curvature at A must be the same as that at A_1, and so forth, for all of the points on or inside of the two triangles. Since the positions of the points A, A_1, etc., are arbitrary, the curvature K must have the same value everywhere on the surface. Thus it is only on a surface with *constant curvature* that we can obtain a geometry with the freedom of movement of Euclidean geometry.

We can now illustrate the different geometries by using different surfaces with constant curvature:

I. Hyperbolic geometry can be illustrated on a surface with constant negative curvature.

II. Euclidean geometry can be illustrated on a surface with constant curvature equal to zero.

III. Elliptic geometry can be illustrated on a surface with constant positive curvature.

I. The simplest surface with constant negative curvature is the so-called *pseudosphere*, which the Italian Eugenio Beltrami used in 1868 to study hyperbolic geometry (Fig. 26). It is a surface of rotation that looks as though one had put together the openings of two bugles.

The plane curve that generates the pseudosphere when it is rotated around the vertical axis in Fig. 26 is called the *tractrix*, or *Huygens'* "dog-curve." The great Dutch physicist Christian Huygens felt that when a reluctant dog is drawn on a leash, his nose describes approximately this curve. From a mathematical point of view it is characterized by the fact that the tangent from an arbitrary point on the curve to the vertical axis has constant length equal to a.

The characteristic feature of hyperbolic geometry can be made clear with the pseudosphere as an illustration. An arbitrary triangle in plane hyperbolic geometry corresponds to a certain geodesic triangle ABC on the pseudosphere in Fig. 26. Since the curvature K is constant we may apply relation (11) of the preceding chapter:

$$A + B + C - \pi = K \times T.$$

The constant K is certainly negative on the pseudosphere. We choose

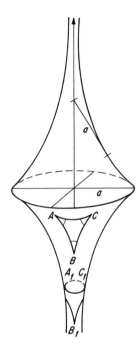

Figure 26
Hyperbolic geometry illustrated on a
pseudosphere.

$K = -1$, which implies that the absolute unit of length mentioned before is equal to 1; then we have

$$T = \pi - (A + B + C).$$

Since the area T of the triangle has a positive value, $(A + B + C)$ is always less than π. This confirms that *the sum of the angles of a hyperbolic
triangle is less than π radians or* 180°.

From the same formula it also follows that the area of a hyperbolic
triangle is always less than or possibly equal to π, which is thus the
area of the largest triangle that can appear in hyperbolic geometry. This
illustrates Gauss's statement in a letter to Wolfgang Bolyai in 1799
(p. 110).

The parallel axiom of hyperbolic geometry is illustrated in Fig. 27, where LM is a geodesic line on the pseudosphere and P a point on the surface that does not lie on LM. The geodesic lines APB and CPD, which meet at a certain angle v, are "parallel" to LM. The area subtended by the angle v is shaded in the figure. None of the geodesic lines through P, which lie in the area subtended by v, intersect LM. One such line is drawn in dashes.

II. The simplest surface with *constant curvature $K = 0$* is the plane. Thus *Euclidean geometry* is most easily illustrated in the plane (rather than on a cylinder or cone). When $K = 0$, relation (11) of the preceding chapter takes on its characteristic form for Euclidean geometry:

$$A + B + C = \pi,$$

which confirms that *the sum of the angles in a Euclidean triangle is always equal to π radians or $180°$*.

III. The simplest surface with *constant positive curvature* is the sphere. *Elliptic geometry* can, after a relatively simple adjustment that we will not go into here, be illustrated on a spherical surface (see Fig. 25). Great circles on the spherical surface correspond to straight lines in the elliptic plane. All of the great circles intersect one another; that is to say, there are no parallel lines. To an arbitrary triangle in plane elliptic geometry there corresponds a certain spherical triangle.

Since K has constant positive value in this case, we can choose $K = +1$, which implies that the absolute unit of length will be equal to 1 and that the radius of the sphere was chosen to be just this length.

Relation (11) now gives the classical formula for the area of the spherical triangle:

$$T = A + B + C - \pi.$$

Since the area of the triangle is positive, $(A + B + C)$ is always larger than π. This confirms that *the sum of the angles in an elliptic triangle is greater than π radians or $180°$*.

It also follows from this formula that the area of the triangle is always finite, and if we compare it with relation (9) of the preceding chapter we see that the angle deviation V is always equal to the area T. This implies that V will only become noticeable when the area of the

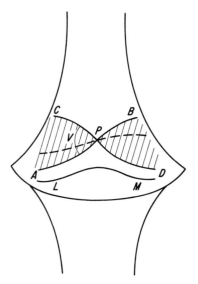

Figure 27
The parallel axiom of hyperbolic
geometry illustrated on a pseudosphere.

triangle is not very small with respect to the total area of the sphere.

The foregoing holds not only for spherical triangles but also for more general curved surfaces with positive curvature, such as a geodesic triangle on the surface of earth. In order to get a measurable value for the angle excess V one must therefore take very large triangles.

Gauss's Measurement of the Triangle
Hohenhagen-Brocken-Inselsberg.

At the end of his paper on curved surfaces published in 1827, Gauss presents the results of the measurement of the record triangle of that time, with vertices on the mountain tops Hohenhagen (H), Brocken (B), and Inselsberg (I). The distances between the points H, B, and I are 69, 85, and 107 kilometers, and the triangle is almost a right triangle.

Using heliotrope signals between the mountain tops, a triangle HBI of light rays was constructed, and the angles H, B, and I were determined. The sum of the angles $H + B + I$ was 180°, in agreement with Euclidean geometry. Then from geodesic measurements Gauss calculated the angles H_1, B_1 and I_1 of the corresponding geodesic triangle on the earth's surface. The sum of these angles exceeded 180° by the insignificant quantity 14.85348″, which thus was the angle excess in this case. It was divided among the different vertices in the following manner

$$H_1 - H = 4.95113''; B_1 - B = 4.95104''; I_1 - I = 4.95131''.$$

The variation around the average of these three numbers depends upon the flattening of the earth toward the north. The deviations are certainly less than 0.0002″, and Gauss concluded the paper with the remark that such small variations can be neglected in all triangles on the earth's surface.

Later mathematicians have suggested that these measurements also had the additional goal of investigating whether or not the triangle *HBI*, constructed of light rays, had a sum of angles that deviated from the Euclidean value of 180°.

The paper on curved surfaces, which to be sure concerned only Euclidean geometry, offers no support for such a suggestion. But in the letter to Taurinus quoted on p. 112, for example, there is a hint in that direction. It is not absolutely out of the question that Gauss, by using his great triangle, attempted to find empirically the deviation from Euclidean geometry in the space of the universe, but there is no direct statement by Gauss in this case. If he had such plans, either for "measurements on earth or in the heavens," he kept them to himself. But from his own remarks we know that he considered geometry an empirical science in the same class as mechanics. The questions which geometry "is right," which one exists "in reality," he thought could be solved through experiment only.

Today there exist many empirical proofs that the geometry that prevails in reality's world is non-Euclidean, although the deviations are so small that we can ignore tham in practical life, for example when we build roads, bridges, or tunnels.

Experimental confirmation has come from atomic physics and from astronomy. Einstein's general theory of relativity gives an expression for the extent to which the universe's real geometry differs from the Euclidean. One can calculate the sum of the angles of a triangle, whose sides are light rays in a gravitational field. The angle deviation for the triangle *HBI* measured by Gauss is so exceedingly small that it lies far beneath the resolution power of the optical instruments of both Gauss's time and our own; the order of magnitude is 10^{-17} second of arc, a decimal expansion that begins with sixteen zeros.

The absolute unit of length in the geometry of the universe may be sought through measurements of cosmic distances and cosmic masses. We do not know the value corresponding to the Gaussian curvature of the non-Euclidean geometry that prevails in the observable universe. We don't even know its "sign." The determination of this universal curvature, which prevails throughout the galactic world, is the most important task of the cosmologists of our day.

The Electromagnetic Telegraph.

In 1831, at Gauss's recommendation, Wilhelm Weber (1804–1891) became professor of physics at Göttingen University. Weber, who was one of the foremost practitioners of experimental physics, ideally complemented the more theoretically inclined Gauss. Together they constituted an unbeatable two-man team in physics. The great difference in age was no hindrance to their friendship. Seen from without, perhaps Weber was not noticed very much in the tremendous shadow cast by Gauss; but on the other hand Gauss always valued him very highly, and after Gauss's death Weber carried out, among other things, experiments that had decisive significance in Maxwell's electromagnetic theory of light.

The most sensational result of Gauss and Weber, from the publicity point of view, was the first electromagnetic telegraph, which began to operate in 1833. To a certain degree it made the two discoverers known even among the lay public.

Gauss was certainly sufficiently famous in earlier days. But his reputation was at first restricted to the researchers in the natural sciences, who at that time did not have the advantage of the publicity of their colleagues of today—one can almost say that they had no publicity at all. The calculation of the orbit of Ceres had extended the waves of Gauss's fame, but they still ended in the academic pool, which at that time was hardly designed for the general public. It was a place where the master of arts was graduated after dizzyingly high dives, or after having been driven around on the surface of the water for a sufficiently long time.

Gauss and Weber's telegraph was built upon Örsted's discovery, in 1820, that an electric current influences a compass needle, and upon

Faraday's discovery, in 1831, of induced currents. The principle is illustrated in the schematic diagram of Fig. 28.

It is a combination of what nowadays are two well-known school experiments. To the right is the transmitter T, which consists of a powerful bar magnet inside a coil. According to Faraday an electrical current is induced in the coil if it is displaced along the bar magnet. This current passes through the telegraph wires to the coil in the receiver R, and according to Örsted's discovery the compass needle NS is deflected.

Gauss and Weber exploited the possibility of the needle swinging in two directions, in the following way:

1. The coil at T is displaced toward the middle of the bar magnet. The compass needle is then deflected in the clockwise direction.

2. The contacts between the coil at T and the telegraph wires are disconnected, the coil is moved to its original position, and the wires are reconnected to the coil with their contacts interchanged. The coil is moved in the same way as before. The compass needle is deflected in the counterclockwise direction.

The deflections were visible on a scale by use of an appropriate light source and a small mirror fastened to the thread from which the needle was hanging. Gauss and Weber could thus transmit two signals, a deflection to the right and a deflection to the left. Figure 29 shows three letters of their alphabet.

The first telegram, which required forty deflections of the compass needle, read: "Michelmann kommt." Michelmann was the university's mechanic who had helped with the installation of the telegraph.

The telegraph wire was about one kilometer long. It ran from the observatory outside Geismar Gate to the old physics institute at Leine Canal. Gauss and Weber exchanged short messages over this telegraph between 1833 and 1845, until one day the wire was destroyed by lightning.

Their apparatus was the first electrical telegraph in practical use. Both Gauss and Weber were aware of its possibilities for further technical development, and among other things they calculated the cost of a telegraph wire to the antipodes. In a letter to Schumacher in 1835

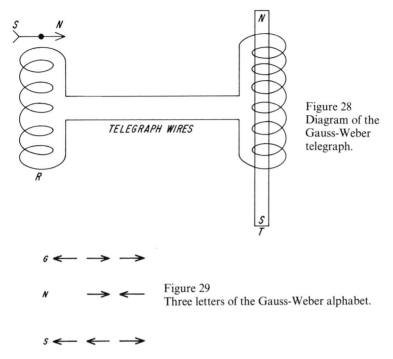

Figure 28
Diagram of the
Gauss-Weber
telegraph.

Figure 29
Three letters of the Gauss-Weber alphabet.

Gauss says that with a larger grant from official sources he thought he would be able to carry out experiments which would bring the electromagnetic telegraph to such a state of completion "that the imagination almost reels."

Most likely Gauss and Weber also saw that their telegraph was the first practical example of the transmission of electrical power over longer distances. In principle the transmitter T is a generator. The hand that moves the coil has now been replaced by, for example, volumes of water that drive a turbine in a power plant. The receiver R can be an electric motor hundreds of kilometers away from the waterfall.

An Absolute System of Measurement.

The result of Gauss's research in magnetism is collected, for the most part, in two works: *Intensitas vis magneticae terrestris ad mensuram absolutam revocata*—Determination of the strength of terrestrial magnetism in absolute units, 1832; and *Allgemeine Theorie des Erdmagnetismus*—General theory of terrestrial magnetism, 1838. Among other things that can be added to these is an "Atlas of terrestrial magnetism" which he published together with Weber in 1840.

The first work is epoch-making in several respects: Through the creation of an absolute system of units, a new principle was introduced for the measurement of quantities in physics; and through Gauss's new methods of observations the earth's magnetic field could be determined with an accuracy heretofore unachieved.

In mechanics one had for a long time used length, mass, and time as the basic units. Gauss proceeded from that point, using the units millimeter, milligram, and second. The novelty consisted in extending the system of units used in mechanics to magnetism (and even to electrostatics) via Coulomb's law. In doing so he found a consistent method of measuring magnetic and electrostatic quantities. Weber later continued the same line of thought in the absolute system of units for electrodynamics.

Gauss and Weber's system was accepted with few changes at an international congress in Paris in 1881. It was the CGS system with centimeter, gram, and second as the basic units. Gauss's name is associated with the units for magnetic field strength and magnetic induction. But physicists have always had difficulty with their units, and many new systems have been tried since then. In the MKSA system now popular (meter, kilogram, second, ampere), a *weber per square meter* is the unit for the magnetic flux density. It is a huge unit that is suitable, for example, for the enormous magnetic fields of cyclotrons. A *gauss* is one ten-thousandth of this unit; it has the same order of magnitude as the flux density (field strength) of the earth's magnetic field at the surface.

In order to determine the magnitude and direction of the earth's magnetic field strength at a point, it is sufficient to determine the declination, inclination, and the horizontal component of magnetism H; the direction of H is the same as that of a compass needle. Gauss concentrated on the measurement of H; his method is used to this day, and it is given in nearly all of the physics textbooks used in high schools.

One first studies the influence that terrestrial magnetism has on a bar magnet with magnetic moment M by letting the bar magnet oscillate in a horizontal plane around its position of equilibrium on the magnetic meridian. From observations of the oscillation time the product MH can be calculated. The quotient M/H is then determined by a deflection experiment, in which the bar magnet deflects another magnet from its state of equilibrium. Finally both H and M are calculated using the product MH and the quotient M/H.

Gauss and Weber not only improved the old methods of measurement but also constructed the bifilar magnetometer, which has the magnet suspended from two threads rather than one, and which can also be used to determine M and H.

Since all magnets are influenced by iron (and certain other metals), one must isolate compass needles and other such things from objects that can give rise to systematic errors. Therefore copper rather than iron was used in the magnetic observatory, which was finished in 1833. Gauss's son Wilhelm was for a while one of the seven assistants, who among other things determined the declination twice a day. Göttingen's magnetic observatory became a center for magnetic measurements, which began to be carried out along the pattern established by Gauss and Weber in many different places, among them Vienna, Kasan, Naples, Dublin, and Uppsala. They also formed a Magnetic Society, of which Alexander von Humboldt was one of the founders.

In his "General theory of terrestrial magnetism" Gauss placed the cause of the earth's magnetic field in the earth's interior, although he later also reckoned with other causes, for example, magnetic disturbances caused by the aurora borealis. This theoretical work is connected with Gauss's research in potential theory, which we shall touch upon in the next chapter.

Geometrical Optics. Capillarity.
The Principle of Least Constraint.

Gauss's activity in astronomy, which of course was his official duty, gave him the impetus to study optics because of the telescopes that he and Harding had ordered. In 1817 Gauss published an article "On the achromatic double objective, especially with respect to more complete removal of chromatic dispersion." Here he combined a convex lens made of crown glass with a concave lens made of flint glass. Many Gauss objectives were ground according to this principle; they were used for a long time, not only in telescopes but also in microscopes. There is also a Gauss ocular that still continues to be used for certain purposes.

Gauss's most important work in optics is "Dioptrische Unter-suchungen"—Investigations on the refraction of light, which was ready in 1840 but was not printed until 1843. His predecessors in this area, for example Euler, Lagrange, and Möbius, had only handled refraction in lenses whose thickness can be neglected, but such a principle often yields a distorted view of reality. Gauss showed, by purely geometrical means, that the refraction of rays near the principle axis in arbitrarily thick or thin lenses or systems of lenses (with a common axis) can be studied using the same simple formulas that hold for a single refracting surface, or a single lens whose thickness can be neglected. The textbook presentation of the fundamental concepts of geometrical optics—focal distance, focal plane, focal point—goes back to Gauss.

Another famous (and lengthy) work, which Gauss wrote in Latin in 1829–1830, is "General principles for a theory of the form of fluids in equilibrium." He began as usual with his predecessor's contributions, and speaks of "sagax Clairaut" and "ill. Laplace," i.e., the sagacious Clairaut and the highly distinguished Laplace.

Gauss then produced a lovely application of the principal radii of curvature for curved surfaces, when he set forth the first fundamental theorem of equilibrium theory. In connection with this he presented, among other things, a method for the determination of the capillarity constant for mercury, using a single drop of that fluid metal.

At the time of his hydrostatics research Gauss wrote a work in German on mechanics: "On a new general principle of mechanics"

(1829). It concerns Gauss's *principle of least constraint*, which he formulates as follows: "The motion of a system consisting of a number of points that arbitrarily influence each other's motion and at the same time are influenced by arbitrary outside conditions, occurs at each moment in greatest possible agreement with free motion, or *under least possible constraint*; the constraint to which the system is subjected at each moment can be measured by the sum of the squares of the deviations of the points from free motion."

Gauss then connects his results to de Maupertuis' principle of "least action" and to d'Alembert's principle. The advantage of Gauss's principle is that instead of using the calculus of variations one may use the methods of differential calculus for determining maxima and minima. His results are believed to have influenced Hertz (who discovered radio waves) in his philosophical investigations of the fundamental problems of mechanics.

After his proof Gauss comments upon the remarkable fact that nature corrects the motion of a body using the same method a mathematician uses in correcting his observations: the method of least squares. "This analogy can be developed farther, but that is not my aim just now," so ends the short metaphysical reflection of the reticent Gauss.

ELEVEN

FUNCTION THEORY AND ARITHMETICAL RESIDUES

Gauss's work in the theory of functions has had less influence on the development of the natural sciences than his contributions to number theory, differential geometry, astronomy, geodesy, magnetism, and optics. But this is not due to the fact that Gauss neglected that area of mathematics. Quite the contrary: from his early youth until his very old age he systematically investigated the general principles of the theory of "transcendental functions"; he had in mind what we now call elliptic functions. We see here an example of Gauss's ability to let many ideas or lines of thought stream through his consciousness on different levels simultaneously and over a long period of time, and to fish one or another of them up to the surface when necessary.

In a letter to Schumacher dated 1808 Gauss wrote: "In the computation of integrals I have always had little interest in matters that simply follow from substitutions, transformations, etc.—in short, making use of a certain mechanism in an appropriate way to transform integrals into algebraic, logarithmic, or circular functions; instead, my real interest has been a more careful and deep consideration of transcendental functions that cannot be transformed into those named above. We can now deal with logarithmic and circular functions as we can with 1 times 1, but that lovely goldmine that contains the higher functions is still almost completely terra incognita. I have worked hard and for a long time in this area, and sometime I shall write a large work on these functions, as I hinted in my Disquiss. aritm. p. 539, Art. 335. One stands in awe before the overflowing treasure of new and highly interesting truths and relations which these functions offer (among them are the formulas that are connected with the rectification of ellipses and hyperbolas)." To rectify is a technical term meaning to straighten or to calculate the length.

Abel and Jacobi.

The terra incognita—unknown land—which Gauss mentioned here is the theory of *elliptic functions*, which got their name from their relation to the problem of calculating the length of an elliptic arc. By "logarithmic and circular functions" Gauss meant a part of what are now called "elementary functions."

The Norwegian Niels Henrik Abel (1802–1829) and the German Carl Gustaf Jacobi (1804–1851) perceived the famous hint in paragraph 335 of Gauss's Arithmetical Research, and started one of the most celebrated races of mathematics. Later investigations have shown that in this case Abel was far superior to his great opponent.

Both Abel and Jacobi had reason to complain of Gauss's indifference to their truly brilliant achievements. Jacobi lived sufficiently long to become famous anyway, perhaps primarily because of Legendre's generous recognition. But Jacobi is thought to never have come in close contact with Gauss. At the great banquet in connection with Gauss's doctoral jubilee in 1849 he sat in the place of honor at Gauss's side, but he was ignored when he tried to talk mathematics. Perhaps the occasion was not very well chosen, in spite of the fact that Gauss was a bit less reserved due to "several glasses of sweet wine," but the result is supposed to have been the same in similar situations. In a letter to his brother concerning the banquet, Jacobi says: "You know that in twenty years he [Gauss] has quoted neither me nor Dirichlet. . . . It is no longer easy to get into a scientific discussion with Gauss; he tries to avoid it by discussing the most uninteresting things in a continuous stream."

Abel's destiny was tragic: it met the three classical criteria of genius, poverty, and contemporary lack of understanding. When he was twenty-one years old he settled one of the great questions of mathematics, with his proof of the impossibility of solving the general fifth-degree algebraic equation by applying only the four basic laws of calculation or taking roots a finite number of times. The method had succeeded for Euclid when applied to equations of the first or second degree, and it had succeeded for Italian mathematicians during the 1500's when applied to equations of third and fourth degree. But then the development had ended for several hundred years in spite of great effort, and in spite of the great progress of mathematics in other areas.

Little by little the fifth-degree equation became a problem in the same class as squaring the circle, trisecting the angle, or doubling the cube. All of the great mathematicians found themselves swamped by papers with "solutions" of the fifth-degree equation, which then proved to be no solution at all. Abel published his proof of the impossibility in 1824 in a little "mémoire" of scarcely six pages, which was printed by the publisher Gröndahl in Oslo (and which is now one of the greatest rarities in the mathematics book market). In the introduction Abel expresses the optimistic hope that "mathematicians might accept this paper with kindliness"; he had thought it would be his letter of introduction on the continent when he shortly set out on his great tour as a Norwegian state stipendiary.

But his well-motivated action became a complete fiasco, since the mathematicians who received the mémoire, after a few glances at the presentation, which was highly compressed for financial reasons, probably threw the thin little booklet in the wastebasket or laid it aside in their own stacks of books and papers. The same tragic reception awaited Abel's later masterpiece on transcendental functions, which he submitted to the French Academy of Science. Cauchy took the paper home but did not read it; he was sufficiently busy with his own epoch-making work. The bulky manuscript vanished in Cauchy's home and was not uncovered again until after Abel's death. After a great deal of trouble it was printed in 1841, but the original vanished again in connection with the printing, and it was not found again until the 1950's, in Florence.

Gauss, like Cauchy, was fully occupied with his own research. He left Abel's mémoire on the fifth-degree equation unread, and left Abel's other work without public comment. As usual he was satisfied to assert the merits of the work on the one hand, and his own priority rights on the other, in letters to his friends. When Abel published his first paper on elliptic functions in "Crelle's Journal" in the fall of 1827, Gauss wrote to Bessel several months later: "On the first hand I have not been able to complete a compilation of my own research on the elliptic functions, carried out since 1798, since I must clear away many other matters beforehand. Mr. Abel has, as I have seen, overtaken me and freed me of about one-third of that labor, especially since he makes all of his development with elegance and clarity. He has chosen precisely the same path that I set out upon in 1798, so the great agreement in the results is not to be wondered at. But to my surprise this agreement extends to and includes the form, and part of the time the choice of symbols also, so that many of his formulas seem to be *pure copies* of my own. In order to prevent any misunderstanding I will also point out that I cannot recall ever having talked with anyone about these matters."

As we see, this letter has a striking resemblance to the letter written to Wolfgang Bolyai in 1832, in which Gauss appraises Johann's non-Euclidean geometry (see p. 114). Gauss is privately appreciative but he again protects his right to priority and to peace and quiet for his work. In both of these cases it is hard to condemn him, but it is equally hard to admire him. His behavior is marked too much by the icy coldness of the Olympian heights where human emotions are frozen away.

When Abel set out upon his European tour his primary cause of hope was Gauss, whom he admired very much. But Gauss was inaccessible, and so with increasing bitterness he circled around Göttingen on his travels to Paris and Berlin.

They never met. Abel died of pulmonary consumption when he was but twenty-six years old. Without question even Gauss would have found a personal meeting with Abel profitable.

The Arithmetic-Geometric Average. Elliptic Functions.

Now we come to a part of Gauss's work in function theory and number theory, which for the most part was found in his papers after his death, and which was then published in his collected works. Dating of the work has been made possible mainly by his diary and his letters.

Gauss wrote Schumacher that he had worked with the arithmetic-geometric average as early as 1791, when he was only fourteen years old. The investigation described below was carried out about 1800. Gauss started with two arbitrary positive numbers a and b, and assumed that a is larger than b, that is, $a > b > 0$. He then constructed two sequences of numbers in the following way.

$$a_0 = a \qquad\qquad b_0 = b$$
$$a_1 = (a+b)/2 \qquad\qquad b_1 = \sqrt{ab}$$
$$a_2 = (a_1+b_1)/2 \qquad\qquad b_2 = \sqrt{a_1 b_1}$$
$$a_3 = (a_2+b_2)/2 \qquad\qquad b_3 = \sqrt{a_2 b_2}$$
$$\dots\dots\dots\dots \qquad\qquad \dots\dots\dots\dots$$
$$\dots\dots\dots\dots \qquad\qquad \dots\dots\dots\dots$$

They can be collected in the recursion formulas

$$a_n = (a_{n-1}+b_{n-1})/2\,;\; b_n = \sqrt{a_{n-1}b_{n-1}},$$

where $n = 1, 2, 3 \dots$, $a_0 = a$, and $b_0 = b$.

Thus the numbers a_n are constructed as the arithmetic averages, and the numbers b_n as the geometric averages, where the index n runs through the natural numbers.

Then Gauss gave a correct proof that these sequences of numbers are convergent, that is to say, that they approach a fixed finite limiting value as the number n grows without bound. At the same time he showed that the limiting value is the same for each of the two sequences, and he called it the *arithmetic-geometric average* $M(a,b)$ of the two numbers a and b.

As usual Gauss excelled at numerical calculations—likely they were made before the exact proof, perhaps when he was exercising his calculating ability as a fourteen-year-old. He studied the cases

$$a = 1, b = 0.2; a = 1, b = 0.6; a = 1, b = 0.8; a = \sqrt{2}, b = 1,$$

and usually carried out the calculations up to and including the twenty-first decimal. Here is a sample from Gauss's tables, in which we have only shown the first ten decimals.

$a\ = 1$	$b\ = 0.2$
$a_1 = 0.6$	$b_1 = 0.44721\ 35954$
$a_2 = 0.52360\ 67977$	$b_2 = 0.51800\ 40128$
$a_3 = 0.52080\ 54052$	$b_3 = 0.52079\ 78709$
$a_4 = 0.52080\ 16381$	$b_4 = 0.52080\ 16380$
$a_5 = 0.52080\ 16381$	$b_5 = 0.52080\ 16381$

By using what is called the method of undetermined coefficients he then, among other things, expanded the reciprocal value of $M(1+x, 1-x)$ in a power series in x; he did not question the existence of such an expansion at that time. The result is

$$\frac{1}{M(1+x,1-x)} = 1 + \left(\frac{1}{2}\right)^2 x^2 + \left(\frac{1\times 3}{2\times 4}\right)^2 x^4 + \left(\frac{1\times 3\times 5}{2\times 4\times 6}\right)^2 x^6 + \ldots$$

He then derived this relation twice and found that it satisfied a certain differential equation, whose general solution he then found. Finally Gauss demonstrated that the complete elliptic integral

$$\frac{1}{\pi}\int_0^\pi \frac{d\phi}{\sqrt{1 - x^2 \cos^2\phi}},$$

which got its name because it first showed up in computing the circumference of an ellipse, has a series expansion which is identical with the one above, that is to say,

$$\frac{1}{M(1-x, 1+x)} = \frac{1}{\pi}\int_0^\pi \frac{d\phi}{\sqrt{1 - x^2 \cos^2\phi}}.$$

This result is to be found in the papers Gauss left behind. In the astronomical results of 1818 mentioned earlier (cf. p. 66) he came to the same conclusion by a different route. He also pointed out there that

Figure 30
The lemniscate.

he had found his methods for calculating elliptic integrals much earlier, independent of Lagrange's and Laplace's investigations which were of a similar character. The article of 1818 on secular perturbations contains the only result about elliptic integrals that Gauss ever published himself.

The starting point for Gauss's work on elliptic functions was the *lemniscate functions*, which are not derived from the ellipse but from the lemniscate, a legacy from the Greeks (Fig. 30). While the ellipse is characterized as the collection of all points for which *the sum* of the distances to two given points is constant, the group of curves to which the lemniscate belongs is characterized by the condition that *the product* of the distances to two given points is constant.

If one wishes to calculate the length y of an arc of a lemniscate, one must consider an integral of the form

$$y = \int_0^x \frac{dt}{\sqrt{1-t^4}}.$$

It is this integral (in indefinite form) that was mentioned earlier as having been referred to in paragraph 335 of Arithmetical Research, and which much later put Abel and Jacobi on the trail of elliptic functions. Gauss says in this paragraph that the same principles that he used in the division of the circle can also be applied, for example, to those transcendental functions that depend on the above-named integral, but that is a task for still another extensive work.

In two diary entries of March 1797, Gauss states that he divided the lemniscate into five equal parts with compass and straightedge, and that the general problem of dividing it into n equal parts leads to an equation of degree n^2. The solution shows how deeply Gauss had penetrated the theory of elliptic functions, even at that early time.

The expression for the length of an arc of a lemniscate is an example of an *elliptic integral*; they had been studied extensively before Gauss's time, principally by Euler and Legendre. But the *elliptic functions* are another matter, and the relation between the two comes about through a notion that is now called *the inverse function* or *the inverse of a given function*.

We shall illustrate this with a simple example. If we assume that x is positive, we can find the inverse function to $y = x^2$, where x is called the independent variable and y the dependent variable, in the following way: We solve the relation $y = x^2$ for x and obtain $x = \sqrt{y}$. No minus sign appears in front of the radical because x is positive. Since one usually denotes the independent variable by x, we will let x and y change places, and in that way we obtain $y = \sqrt{x}$. The two functions $y = x^2$ and $y = \sqrt{x}$ are then inverses to each other for positive values of x.

Other examples of inverse functions among the elementary functions are given by the pair $y = \log_e x$ and $y = e^x$, where e is the base of the natural logarithm system; and by the pair $y = \sin x$ and $y = \text{arc} \sin x$, the last of which is read *the inverse sine of x*.

It is true in general that if we can solve the one-to-one relation $y = f(x)$ for x as a single-valued function of y, and if in so doing we obtain $x = g(y)$, then the second function is the inverse of the first, and conversely. For practical reasons we then exchange x and y and obtain the inverse in the form $y = g(x)$.

Now we return to the elliptic integral

$$y = \int_0^x \frac{dt}{\sqrt{1-t^4}}.$$

It expresses a relation between y and x of the type $y = f(x)$. But in this case we cannot express x as a function of y, that is to say, $x = g(y)$, by using elementary functions. When such a situation comes about in mathematics, it is solved quite simple by introducing a special symbol for the new function. In this case Gauss introduced the notation

$$x = \sin \text{lemn } y$$

for the inverse of the elliptic integral on the preceding page. After x and y change places it becomes

$$y = \text{sin lemn } x,$$

which is read *the sine of the lemniscate of x*. He also introduced cos lemn x, that is to say, *the cosine of the lemniscate of x*.

These are the Gaussian lemniscate functions, the first elliptic functions. The name comes from the fact that Gauss developed his theory in analogy with the theory of the trigonometric functions sin x and cos x. He found many formulas for his new functions which are analogous to the usual trigonometric formulas. He also found that the lemniscate functions have *two* periods, as distinguished from the trigonometric functions, which have only *one* period.

If one can find a positive number k associated with the function $f(x)$, such that

$$f(x) = f(x+k)$$

for all values of x for which the function is defined, then one says that $f(x)$ has *period k*. For example, for the function $f(x) = \sin x$ we have $\sin x = \sin (x+2\pi)$, that is, sin x has period 2π radians, or $360°$. This is due to the fact that when the side of an angle is rotated by one revolution, the angles are returned to their original position.

In the same way one says that a function $f(x)$ has two periods if one can find two different numbers k_1 and k_2, such that

$$f(x) = f(x+k_1) \text{ and } f(x) = f(x+k_2)$$

for all of the values of x for which the function is defined.

Gauss showed that sin lemn x has not only a real period, which he denoted by 2ω, but also the purely imaginary period $2\omega i$. In a journal entry of May 1797, he records that he had found the relation between ω and the arithmetic-geometric average. The formula is

$$\omega = 2 \int_0^1 \frac{dt}{\sqrt{1-t^4}} = \frac{\pi}{M(1, \sqrt{2})},$$

and Gauss first found it in a purely empirical way by checking the agreement between the two sides out to the eleventh decimal. He proved it at the end of 1799.

The lemniscate functions belong to the larger class of elliptic functions. If $P(t)$ is a polynomial of third or fourth degree, then an integral of the form

$$y = \int_0^x \frac{dt}{\sqrt{P(t)}}$$

is called an *elliptic integral*. The inverse is a general elliptic function. If, for example, we choose

$$P(t) = 1 - t^4,$$

we obtain the sine of the lemniscate.

Also among Gauss's great contributions to this area is his fundamental investigation of what are called modular functions (cf. p. 38). Here he also used the results he had achieved in his studies of binary quadratic forms (cf. p. 58).

Gauss's investigations of the elliptic functions are perhaps the best proof of the universal scope of his mathematical ability. For here three widely differing elements—the arithmetic-geometric average, the lemniscate, and the quadratic form—undergo a transmutation, and out of this alchemistical process comes one of the heretofore unknown elements of mathematics: the elliptic function. In accordance with his principles, Gauss likely wanted to give a complete presentation of the entire theory of elliptic functions to the general public or else none at all. But for a complete treatment of the subject the results obtainable through integration in the complex plane and through the introduction of Riemann surfaces were lacking.

Gauss made no public statement concerning the elliptic functions; but in 1818, at the earliest, he made a collection of his results under the title "One hundred theorems about the new transcendental functions," which probably was intended for publication.

Analytic Functions. Infinite Series. Potential Theory.

We have already mentioned complex numbers and the complex or Gaussian plane. In the same way that one constructs functions $f(x)$ of a *real* variable x, one can also construct functions $f(z)$ of a *complex* variable $z = x + iy$. Simple examples are $f(z) = 2z + 5$ and $f(z) = z^2$. For $z = 1 + 2i$ the latter function has the value

$$f(1+2i) = (1+2i)^2 = 1 + 4i + 4i^2 = -3 + 4i.$$

Derivatives and integrals of functions in the complex plane are defined in analogy with corresponding definitions for functions of a real variable. If a function $f(z)$ has a derivative at each point of a region D in the complex plane, one says that $f(z)$ is an *analytic function* on the region D.

Integration in the complex plane generally occurs along open or closed curves. Here we shall only mention the results in this area that Gauss communicated to Bessel in a letter of December 1811. From the style of writing it is apparent that at that time he had completely mastered the geometric representation of the complex numbers (cf. p. 26). He investigates the integral of a complex function along a curve in the complex plane. He explains the many-valuedness of certain functions that are presented with the help of integrals by considering curves that make one or more revolutions about the singular points (poles). As an example he chose log z, where z is a complex number, and he explains the infinitely many branches of this function with the help of the formula

$$\int_C \frac{dz}{z} = 2k\pi i,$$

where the path of integration is a curve C that travels around the origin k times. If $k = 2$, for example, the curve makes two revolutions around the origin in the positive direction. In Fig. 31 we have $k = 1$. If $k = -2$ then the curve makes two revolutions about the origin in the opposite direction.

In the same letter he formulates—and clearly he also had a proof for—the fundamental theorem in the theory of functions of a complex variable. Independent of Gauss, the great French mathematician Augustin Cauchy (1789–1857) published the same theorem in 1825 in his theory of analytic functions. With perfect justification it is called Cauchy's integral theorem.

It is likely that Gauss intended to apply the results mentioned here, at least to the looming exposition of elliptic functions.

Gauss used infinite sequences and series in his daily work, not only in mathematics but in astronomy, geodesy, and physics; he mastered the techniques of expansion in series at an early age. He also thought about the theoretical formulation of the notion of limiting value. In an unfinished article written around 1800, "Fundamental concepts in the principles of series," he formulated the notion of the limit of a sequence in a fashion far ahead of the times.

Gauss introduced there the notions of upper bound and least upper bound G; he also introduced the notions of lower bound and greatest lower bound g. Furthermore he introduced the "final upper bound" H and the "final lower bound" h. If $H = h$, then their common value was called the *absolute limit* (limiting value) of the sequence. Gauss's definitions nearly agree with the present-day definitions of upper bound G, lower bound g, limit superior H, limit inferior h, and the condition $H = h$ for the existence of the limiting value.

In a much later handwritten manuscript—after 1831—he returns

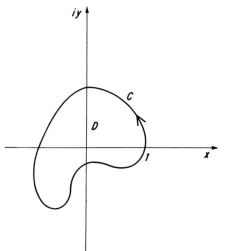

Figure 31
In the complex plane one
often integrates over a
closed curve C about the
region D.

again to an old-fashioned unclear terminology and unclear limit concept, but that did not hinder his scrupulous treatment of series expansions.

Gauss's most important work on infinite series, published in 1813, is "*General investigations of the infinite series*

$$1 + \frac{a \times b}{1 \times c}x + \frac{a(a+1)b(b+1)}{1 \times 2 \times c(c+1)}x^2 + \frac{a(a+1)(a+2)b(b+1)(b+2)}{1 \times 2 \times 3 \times c(c+1)(c+2)}x^3 + \ldots"$$

This series is called the *hypergeometric series*. It is the Proteus of mathematics, since by giving different values to the constants a, b, and c, or by transforming x, one can obtain the series expansions of most of the functions that commonly arise in analysis. For $a = 1$ and $b = c$ we obtain the geometric series

$$1 + x + x^2 + x^3 + \ldots$$

If $a = b = 1/2$ and $c = 1$, and if we replace x with x^2, we obtain the series expansion for the reciprocal value of $M(1+x, 1-x)$, that was dealt with by Gauss earlier. There are many other examples of this sort.

The paper has a purely analytic character, but it also contains many series expansions, and the relations that Gauss derived from his perturbation theory for the planets. From a logical viewpoint it is remarkable in that it contains precise criteria for the convergence of a power series. Except for his doctoral thesis it is Gauss's most important contribution as a rigorist.

The hypergeometric series had been introduced by Euler in a work of 1769. Gauss used it theoretically, part of the time, and created the theory of hypergeometric functions; he used it in a purely practical fashion, at other times, for the solutions of problems in mathematics, astronomy, physics, and mechanics.

The paper of 1813 was called "Part One." Gauss had thought of publishing two more parts. The second part was to present a theory of those differential equations for which the hypergeometric series is what is called a particular solution, and the third part was to contain the theory of the general elliptic functions.

These latter parts were never written and it was only when people dug through the papers left behind by Gauss that they unearthed, among other things, those theorems which will forever bear the names of Abel and Jacobi, but which Gauss had discovered long before they were born.

Gauss's work in *potential theory* stands on the boundary between pure and applied mathematics; on one side it has given impetus to a branch of research in pure mathematics, on the other side it has had widespread application in physics. The term *potential* itself goes back to Gauss, who introduced it in a great work of 1839. Investigations in the same field had been carried out earlier, by the Englishman George Green, among others, in his work of 1828; the term potential function appears there but it is used in a different sense than that of Gauss's potential. For a long time Green's paper was virtually unknown, and certainly Gauss, as usual, worked independently of all others.

The final conclusion about Gauss's work in function theory is that he let the deepest and most far-reaching results remain within the walls of his own workroom. He never found time for the large and beautiful synthesis of the "transcendental functions" that he dreamed about.

Therefore his results were rediscovered by Abel, Jacobi, Cauchy, and others, and were incorporated into mathematics independent of him.

Biquadratic and Cubic Residues. Fermat's Great Theorem.

In a planned eighth section of Arithmetical Research Gauss intended to handle congruences of degree higher than two, and to set up corresponding reciprocity laws (cf. p. 56). Thus he was concerned with the solvability or unsolvability of

$$x^n \equiv q \pmod{p} \text{ and } x^n \equiv p \pmod{q},$$

where p and q are different prime numbers. He wanted first of all to consider the cases $n = 3$ and $n = 4$, the cubic and biquadratic residues.

For various reasons, finances for one, this eighth section was put aside, and he did not return to the subject until 1825, in a paper that was printed in 1828: "*Theoria residuorum biquadraticorum. Commentatio prima*"—Theory of biquadratic residues. First section. (A second section came out in 1831–1832.)

This was a difficult problem even for Gauss, and only after many false starts did he discover the surprising solution: The usual whole numbers are adequate for the law of quadratic reciprocity, but they are *not* adequate for formulating the law of biquadratic reciprocity; here one must introduce a new type of whole number. This is the *Gaussian integer*, a complex number of the form $a + ib$, where a and b are *whole numbers*.

In this paper Gauss opened a whole new area of number theory and algebra, and in it he also gave his own public approval to imaginary numbers. It is believed that he investigated cubic reciprocity at the same time, but he never published a thing on that subject. It was solved by

Gotthold Eisenstein (1823–1852), whom Gauss valued highest among his disciples.

Before we leave Gauss's mathematical contributions we shall mention his attitude toward *Fermat's great theorem*, since this problem draws much attention even outside the circle of professional mathematicians. If we first study the Diophantine equation (cf. p. 53) of second degree,

$$x^2 + y^2 = z^2,$$

the problem involved is to find the whole numbers x, y, and z that satisfy the equation. In this case there exist infinitely many solutions, a fact that was known to the geometers of antiquity; for example, 3, 4, and 5 is one solution, as is 7, 24, and 25. When the solution consists of natural numbers, as in the examples, they are usually called pythagorean numbers, since they can be used as the lengths of the sides of a right triangle.

If we increase the degree of the diophantine equation, we obtain in turn

$$x^3 + y^3 = z^3$$
$$x^4 + y^4 = z^4$$
$$\cdots\cdots\cdots\cdots$$
$$\cdots\cdots\cdots\cdots$$
$$x^n + y^n = z^n.$$

Fermat's so-called great theorem states that the diophantine equation

$$x^n + y^n = z^n,$$

where n is a natural number larger than 2, is *not* solvable in whole numbers x, y, and z, all of which are different from zero.

Fermat formulated this theorem as an entry in the margin of his copy of Diophantos' *Arithmetica*. At the same time he wrote that he had found a truly wonderful proof, but that the margin was too cramped to contain it.

He did not communicate his proof anywhere else. According to one version, which certainly belongs to the world of myth, he is said to have written it down on his shirt-cuff, perhaps during a long-winded deliberation before the court in Toulouse, where he was a member of the magistrate's court. But he was an absent-minded man, and the remarkable proof was obliterated in the wash. In the case $n = 4$, however, we know that Fermat actually proved his theorem. As for the rest, he was such a great mathematician that his entry in the margin must be taken seriously, even if we can never know if his proof was correct. In spite of the fact that all of the results obtained so far indicate that the theorem is valid, one still has not been able to find a proof that covers all values of n.

Gauss did not set any great value on Fermat's great theorem. The French Academy of Science had posed the proof or disproof of the theorem as a prize problem during the period 1816–1818. Olbers, in a letter, urged Gauss to take part in the competition. In March of 1816 Gauss answered: "I am quite obliged for your report of the Paris prize. But I must say that Fermat's theorem, considered as an isolated proposition, interests me very little; I could very easily propose a whole string of such propositions, which no one should be able to prove or use."

It isn't known to what propositions Gauss is alluding here. Perhaps there was no one who dared ask him.

TWELVE PERSONAL FACTS ABOUT GAUSS

His Children.

Gauss's children have appeared here and there in the foregoing presentation; now we shall say a bit more about them.

Joseph, who was born in 1806, was the first child in Gauss's marriage with Johanna Osthoff. He had a good middle-class career that began in his father's shadow with his work on the triangulation between Göttingen and Altona. He then became an artillery officer and one of the leaders of the continued triangulation measurements of the whole of Hannover. In 1836 he traveled to the United States to study new methods being used in the construction of railroads. He is thought not to have been particularly fond of the military life, and after twenty years service he resigned from the corps of engineers as a first lieutenant, in 1846. At about the same time he became one of the four directors of Hannover's railroad system. Ten years later he received the title "Oberbaurath." Joseph resembled his father both in appearance and in certain traits of character. He died in 1873.

Wilhemina, born in 1808, was the second child in Gauss's first marriage. She was called Minna and very much resembled her mother both in appearance and in temperament. Gauss asserted that she was the very image of her mother, and she became his favorite among the children. In 1830 she married Heinrich Ewald, who was professor of theology and oriental languages at Göttingen. He was known, among other things, for an absentmindedness that was more than common, even in academic circles, but the marriage was looked upon as a happy one. Minna's resemblance to her mother also included her frail health, unfortunately. She died in 1840, probably of pulmonary consumption.

Ludwig, born in 1809, was the third child in Gauss's first marriage.

He was "der arme kleine Louis," who died at only six months of age. His birth had cost his mother her life.

Not ten full months after his dearly beloved Johanna died, Gauss married Minna Waldeck, who came from a wealthy family. She also had fragile health. She sought remedy through cures at the spas of Pyrmont and Ems, but that period's weapon against her illness, which probably was also pulmonary consumption, was not effective, and she died in 1831. Three children were born in the second marriage, Eugene in 1811, Wilhelm in 1813, and Therese in 1816.

Eugene merits special attention, because he among all of the grown children of Gauss had the most of Gauss's ability, both in mathematics and languages. When he finished school he wanted to study philology at the university. But Gauss opposed that with all of his fatherly and scientific authority. Eugene could not withstand such crushing opposition. He followed his father's "advice" and signed up as a student in the law faculty.

If Eugene had been of a passive disposition perhaps he could have continued and become a reasonably good jurist. But he was of an unsettled nature and had a violent temper. He looked into playing cards and beer glasses more than he did law books. This was primarily a protest against fatherly authority, and it is supposed to have led quickly to what was known as an "old man's debt," in which the father had to clear up his son's obligations. To counter his father's sharp reprimands Eugene put forth his intention to travel to the United States. The outcome was that he set off from home without packing and without saying good-bye to anyone. Gauss believed that he was in Hamburg, and in a letter to Schumacher he asked him to arrange with Hamburg's chief of police to have him found. But Eugene had gone to Bremen, and Gauss followed him there; he was prepared for all possibilities, having brought both travel money and a trunk for his son. They met at Olbers', but Eugene did not care to partake of the prodigal son's meal. Father and son parted—never to meet again—and in December, 1830, Eugene stepped ashore in New York.

After a couple of months he enlisted in the U.S. Army and advanced to the rank of sergeant. He was not happy as a soldier and in a letter to his father asked for help in obtaining his discharge from the army; he had requested a discharge due to his near sightedness but the request was denied. In reply he received a long moral lecture, filled with reprimands and hopes of future improvement. In a letter to Olbers in June of 1831 Gauss attributes his own uneasiness to the "good-for-nothing" (Taugenichts) in America.

But his misgivings were not justified. During his military service Eugene became a pious and zealous Presbyterian, and he even wanted to become a missionary. After his discharge from the army he took a position with a fur company that had dealings on the Mississippi–Missouri. Among other places, he served in an office on the Wisconsin prairie. There he readily learned the language of the Sioux and helped a missionary establish an alphabet for a bible translation. He also knew Greek.

Toward the end of the 1830's Gauss became convinced that Eugene had improved and sent him the proceeds of his inheritance from his mother, which had been completely blocked until then by a codicil to the will concerning good behavior. In 1840 Eugene settled down as a businessman in St. Charles, Missouri. Among other things he was a grain merchant and a lumber dealer. He organized the First National Bank and became its first director. He married, had seven children, and lived in a large stone house with a surrounding park. In 1885 he moved to his country estate in Boone County, outside Columbia, Missouri, and in 1896 he died there, the last of Gauss's children.

The relationship between father and son was softened by the years and by success, but it is thought never to have been completely restored. Eugene burned most of his letters from his father. A gold medal, awarded to Gauss by King George V, was inherited by Eugene and melted down to make frames for his eyeglasses.

In spite of his gold-medaled dioptrics Eugene gradually became blind. When he was over eighty years old he used to divert himself by calculating in his head. Among other things he calculated the sum to which one dollar would have grown at four percent interest, beginning at the time of Adam's birth—which he took, according to his theological authorities, to be 6,000 years before 1894. The amount, expressed in volume of gold, yields a cube in which the presently observable universe vanishes like a drop of water in a bathtube. Eugene calculated with common arithmetic and his only help was his son Theodore, who, every few days, took down his father's figures in dictation. For Eugene did not really trust his memory. But he could have safely done so, for on at least one occasion he had to correct Theodore, who had calculated on his own and erred. It became apparent that the blind eighty-three-year-old Eugene could retain a string of thirty digits intact in his memory for several days. His result was checked later by a professor of astronomy.

A man with this calculating ability, who in addition could speak the languages of both the Greek heroes of mythology and the Sioux chiefs, was certainly a very gifted man of whom Papa Gauss need not have been ashamed at home in Göttingen.

In broad outline Wilhelm followed the same route as Eugene, but he never attempted to rebel against his father. After finishing school he devoted himself to farming, with the intention of eventually owning his own farm. After his apprenticeship he held administrative posts on many different estates, but he was not happy in his work, and he became a businessman in Potsdam for two years.

In 1837 he married a niece of his father's good friend Bessel. He had long had plans to emigrate to the United States, and with his father's blessing the newlyweds journeyed to New Orleans toward the end of 1837. Quite soon Wilhelm was able to buy his own farm in the vicinity of St. Charles. But he had bad luck both in farming and in health, and so he again set out upon a business life. There he had better luck, and after another interlude as a farmer he established himself as a wholesale shoe dealer in St. Louis. He had eight children and died a millionaire in 1879.

Therese resembled her mother very much both in temperament and appearance. When she was about twenty years old she took over Gauss's household. Earlier she had cared for her grandmother Dorothea, who had lived at the observatory since it was completed and who was always the object of Gauss's filial piety. After Gauss's mother died at ninety-seven, Therese completely dedicated her life to her father. After his death she married an actor and theatrical producer, with whom she had exchanged letters for fourteen years. She died without children in 1864.

Gauss was a tender father only toward his daughters—although one may certainly wonder how he would have reacted if they had shown any signs of emancipation. He handled his sons with a harshness which was basically the same type of harshness that he himself experienced from his father. He was not as unrestrained as his father, but both of them perceived fatherly authority as a moral parallel axiom, whose obviousness was not to be questioned. They were certainly not alone in that, since it is indeed true that only in our time has non-Euclidean geometry been accepted in that area. It is uncertain whether or not Gauss ever became aware of the fact that he treated Eugene worse than his own father had one time treated him.

Gauss as Seen Outside Mathematics.

Gauss's political views were strongly conservative and intellectual—aristocratic, the students quite correctly regarded him as a reactionary. In 1837 King Ernst August, the fifth son of King George III of England, nullified the liberal constitution of 1833 for Hannover. That step brought forth, among other things, a public protest from seven professors at Göttingen, among them Ewald, Weber, and the Grimm brothers. They were all dismissed from their positions, three of them were banished from the land, and the others were allowed to remain in the country only if they "behaved themselves well." Göttingen was occupied by the military for a while.

Gauss's name, which would have given the protest quite a different weight, was lacking. He did not like revolt against authority and he wished, above all else, to protect his peace and quiet for his work. In any case he asked the influential Alexander von Humboldt to speak up for Weber, but not for Ewald, since he did not want to be suspected of subjective viewpoints with regard to his son-in-law. But Weber received no help from that appeal. At a banquet the crude Ernst August communicated his view of the matter to Humboldt: "With my money I can buy as many ballet dancers, whores, and professors as I wish."

The February revolution of 1848 was thoroughly disapproved of by Gauss, even though Weber and Ewald were reappointed to their positions thanks to the resurgence of the liberal ideas. Perhaps he was afraid that events in the style of the Napoleonic wars would again sweep over Germany. In Göttingen, in 1848, a literary club called "Against the Revolution" was formed. Gauss became a member and sat in the club's reading room every day between eleven and one. He was an ardent newspaper and magazine reader. He used to gather into a pile everything around that he wanted to look into. He would order them chronologically and sit down upon them, then haul them forth one after another and read through the different papers. He wrote down especially interesting things in a little notebook. If a student happened to be reading a newspaper that he wanted to have, Gauss would stand and stare at his victim until he blushingly handed it over. The students called him "the newspaper tiger."

Gauss was well at home in German literature and philosophy. His favorite author was Jean Paul, pseudonym for Friedrich Richter, who wrote the best-selling novels of his time in a romantic style. Gauss considered him "incomparable" (unvergleichlich).

On the other hand he did not place Goethe very high although he read all of his writings. They did not correspond and they never met; this is a pity, since those two divine councillors would have had a good many thoughts to exchange, in any event many concerning the theory of colors.

Gauss considered Schiller's philosophy repulsive and considered certain of his poems blasphemous. On the other hand he referred to "Archimedes und der Schüler" in his inauguration lecture. Among the dramas he liked "Wallenstein" best.

Gauss's recreational reading of foreign literature was facilitated by his gift for languages. He read Russian and Danish—there were seventy-five Russian volumes in his library, among them eight volumes of Pushkin, and the collected writings of Holberg in the original language—whereas his knowledge of Italian, Spanish, and Swedish was superficial. He knew the classic Greek and Latin authors thoroughly. He himself wrote a fluent and perfectly formed Latin in his writings; perhaps there is some psychoanalytic explanation for the fact that he even used Latin in his secret diary. Of the French classics he read, for example, Montaigne, Rousseau, Voltaire, Montesquieu, and Boileau.

The living foreign language that Gauss read most and commanded best was English. It is typical of his melancholic disposition that he found it difficult to read Shakespeare's tragedies. He felt there was sufficient sorrow in real life. He was a naive reader who wanted to have a happy ending to the story. But he took one of his mottoes from King Lear:

Nature, thou art mine goddess! To thy laws my services are bound.

It was an obvious motto for a person who saw no difference between what we now call pure and applied mathematics.

Gauss admired Sir Walter Scott very much and read all of his work carefully. To his delight he discovered an astronomical error in a description of natural scenery: "the full moon rose in the northwest." He corrected both his own copy and those of the Göttingen book dealers.

Such a thing afforded him as much joy as finding an error in a table of logarithms.

It is not uncommon for a connection between music and mathematics to exist, but Gauss is thought not to have been particularly musical. He played no instrument and never burst into song. But he liked very much to listen to songs, and he used to write down the text to those *Lieder* that he appreciated most. He went to a concert approximately once every ten years, with the exception of February 2 and 4, 1850, when Jenny Lind was in Göttingen.

In Gauss's serious character there was no great room for playfulness. But he did not completely lack humor. When Göttingen University celebrated its centenary, for instance, he joked about the "deluge of poems" that was let loose by different drivelers, and in a letter to Olbers he gave a satirical description of the endless ceremonies, which was the death of one of his colleagues and nearly the death of himself.

He did not like official pomp and pageantry, and he never concerned himself much with the official marks of honor which rained over him in a continual shower for fifty years. Altogether he accepted about seventy-five official honors of different kinds: memberships in various scientific societies, scientific prizes, diplomas, medals, orders, and so forth. All of Europe took part in this awarding of prizes. In 1818 he became *Hofrat* to King George and in 1845, without murmur, he accepted the title *Geheimer Hofrat* of Ernst August.

He always lived under the same Spartan forms that he had become used to in his childhood years. But he smoked a pipe and enjoyed drinking a glass of wine (of a special French brand) in the company of friends. This simple conduct of life, combined with an ability in economic matters, resulted in his leaving behind a fortune of 153,000 thalers. (The inheritance from Minna Gauss had gone entirely to the children.) A couple of days after his death another 18,000 thalers in cash was found, hidden in his writing desk, closets, and bureau drawers; so far as is known, none was in the mattress. Gauss's basic salary as a professor was 1,000 thalers per year, plus fees from the students. But Gauss always had small attendance in his courses. It is rather striking that under these conditions, without really being a miser, he was able to build up such a fortune. It consisted mostly of investments in different countries, that Gauss had chosen in a very skillful way. The cash made up a reserve with which he did not speculate.

Gauss's attitude toward metaphysical and religious questions was stamped with his logical and natural scientific turn of mind. In common with many great mathematicians, for whom abstract symbols are the true reality, he was not a materialist in the classical meaning of the word, for example, when he said: "For the soul there is a satisfaction of a higher type; the material sort is not at all necessary. Whether I apply mathematics to a few clods of soil, that we call planets, or to purely mathematical problems, is in itself unimportant; but the latter application gives me greater satisfaction."

During the last months of his life he conversed rather often with his colleague Rudolf Wagner, a professor of physiology and zoology, who was interested in spiritual matters. With reference to Fechner's work they discussed the use of mathematics in psychology, in regard to which Gauss said: "I hardly believe there are any data in psychology that can be expressed mathematically. But one cannot know that with certainty until he has made experiments. God alone is in possession of the mathematical foundations of psychic phenomena."

The "principle of least constraint" mentioned earlier is thought to have been, for Gauss, evidence of a general principle that guides the universe in both its mechanical and ethical forms of manifestation. Both this principle and the reply to Wagner seem to refer to a world-spirit, which alone possesses, and is, the mathematical structure of all material and spiritual phenomena. It is a cosmic religiousness that resembles Einstein's.

Gauss did not believe in the Christian dogma word for word, but in his later years he sought to build a belief in life after death, based upon rational argument. He concluded one of his conversations with Wagner with the following words: "Without immortality the whole world should be nonsense, all of creation an absurdity."

He conveyed similar thoughts to Wolfgang Bolyai in a letter of 1848, when he was seventy-one years old. Here it is his loneliness and tragic conception of the emptiness of earthly existence which supports the hope of survival after death: "It is true that my life has been adorned with many things upon which the world looks with envy. But believe me, dear Bolyai, the *bitter* sides of life, in any case those that I have come to know, which pass like a red thread through all the world, and against which one becomes all the more defenseless in old age, are not countered to one one-hundreth part by the joyous. I will gladly concede that the same fate, that has been and still is so hard for me to bear, would have been much easier for many others, but our state of mind belongs to our ego, the creator of our existence has given it to us, and we are little able to change anything about it. On the other hand it makes us conscious of the emptiness of life, which in any case the greatest part of mankind must know when the end is approached, and to me it offers the strongest guarantee that a more beautiful metamorphosis shall follow. With that, dear friend, we must console ourselves, and through it seek to attain the equilibrium which is necessary to endure to the end."

Before Gauss was stricken by sickness he had little time to spare for religion and metaphysics, and the most accurate apprehension of his attitude toward these things probably comes from his observation: "There are questions upon whose answering I should set immeasurably more value than mathematical questions; for example, those concerning ethics, our relation to God, our destiny, and our future; but their solution lies completely beyond our reach and completely outside the area of science."

The Last Years.

Although Gauss was often upset about his health, he was healthy almost all of his life. His capacity for work was colossal and it is best likened to the contributions of different teams of researchers over a period of many years, in mathematics, astronomy, geodesy, and physics. He must have been as strong as a bear in order not to have broken under such a burden. He distrusted all doctors and did not pay much attention to Olbers' warnings. During the winters of 1852 and 1853 the symptoms are thought to have become more serious, and in January of 1854 Gauss underwent a careful examination by his colleague Wilhelm Baum, professor of surgery. The diagnosis was enlargement of the heart, and Baum also said—probably in answer to a direct question—that there was little hope of improvement; instead the symptoms indicated that Gauss would not survive any great length of time.

Toward the end of the year the illness worsened. As a consequence of his failing heart he developed dropsy and increasing difficulty in breathing. In December of 1854 Gauss wrote his will. By then his previously clear hand-writing had failed. On January 5, 1855, however, he wrote a letter to the university officials concerning the repair of his house during the coming spring.

The last days were difficult, but between heart attacks Gauss read a great deal, half lying in an easy chair. Sartorius visited him the middle of January and observed that his clear blue eyes had not lost their gleam. The end came about a month later. In the morning of February 23, Gauss died peacefully in his sleep. He was seventy-seven years old.

The burial on February 26 became an academic funeral ceremony with participants from far outside the university circles. Wreathed in laurel, Gauss lay in state under the observatory's cupola, and the coffin was carried away by twelve students, among them Dedekind. It was placed in the earth in St. Albans cemetery in Göttingen.

The Collected Work. Gauss as Teacher. Summing up.

Many monuments were raised to the memory of Gauss, but his lasting monument is his *Collected Works*. Aside from book reviews and similar notices, Gauss published approximately 150 papers from 1799 until his death. There was also an enormous number of papers left behind in a more or less completed condition. All of this, and even appropriate pieces of Gauss's correspondence, were arranged and annotated under the direction of the Royal Scientific Society of Göttingen. It is an extraordinarily thorough edition in twelve large volumes that was carried out during the years 1863–1933 by ten German scientists, each an expert in his area.

This work also proved Gauss's priority in several areas, as we mentioned earlier. But hardly any other mathematician has had his work so carefully gone over in every little detail as has Gauss. Therefore it is not impossible that a corresponding investigation of other great mathematicians, say for example those from 1750 to 1850, could change the priority conditions in a number of cases. It was a rich century. Wolfgang Bolyai, like Gauss, loved the flowery language of the romantics. He described the situation well, when in 1823 he urged his son Johann to hurry with the publication of his new geometry: ". . . for ideas appear at the same time in different places, like violets in spring." But on the whole the ill-tempered priority arguments, in which Gauss certainly never took any public part, should now be buried in peace.

Gauss was professor of astronomy, and from 1808 to 1854, with few exceptions, he lectured on that subject and its relations to mathematics in the form of orbit determination and perturbation theory, or in the form of error calculation and the method of least squares. His geodesic and magnetic measurements came forth more sporadically but follow the same general scheme. The exceptions are a series of lectures on number theory in the winter of 1809 and a series on the general theory of curved surfaces in the summer of 1827. He also touched upon the application of the calculus of probability to crystallography and possibly upon numerical methods of integration—two areas we did not mention earlier because his contributions were relatively insignificant.

Like a parson who after a couple of years of preaching begins over again, Gauss gave the same lectures year after year with very little variation. He was not a bad lecturer—quite the opposite—but he quite simply was not moved by teaching and thought that it robbed him of his time, which became all the more precious with approaching old age. In addition, he felt that the really gifted students could take care of themselves and only needed a few suggestions now and then. He often returned to that theme in his letters.

As early as 1802 he wrote to Olbers: "I have a real aversion to teaching. For a professor of mathematics it consists of eternal work to just teach the ABC's of his science; most of the few students who go on continue to gather a pile of information and become only half-educated, for the rare gifted students will not allow themselves to be educated through lectures but instead learn by themselves. And through this thankless work the professor loses his precious time."

Of the gifted students, Gauss wrote to Schumacher in 1808: "One does not need to take such a student by the hand and lead him to the goal; one only needs to give him a suggestion now and then, so that he will find the shortest way."

Concerning students without preparation or ability he wrote to Bessel in 1810: "This winter I have two courses for three listeners, of whom only one is moderately prepared, another hardly prepared, while the third lacks both preparation and aptitude."

Gauss's demand that the students be prepared plus his Olympian elevation resulted in his having five to ten students while his colleague Thibaut, who lectured on elementary mathematics, had one hundred. (In 1850 Göttingen had about 10,000 inhabitants and 850 students.)

When Gauss gave his lectures the listeners sat around a table covered with books. Gauss himself sat in an armchair at one of the narrow ends of the table with a rather small blackboard beside him on an easel; this was sufficient for his diminutive calligraphy of numbers and other mathematical symbols. In his older days a black velvet cap almost always concealed his fine head of hair, which slowly became white. He sat in a comfortable position, his head bent slightly forward, his eyes down, and his hands clasped on his knees. He did not like to have the students take notes. If someone did so anyway, he might be interrupted with: "Don't write, follow along more carefully instead."

No one has given a better description of what it was like when genius suddenly stared forth from behind the teacher's humdrum mask than the great German mathematician Richard Dedekind (1831–1916), who was one of the nine students attending Gauss's lectures on the method of least squares during the winter of 1850–1851: "Gauss spoke freely, very clearly, simply and right to the point; but when he wished to put forth a new viewpoint in which he used a special and characteristic choice of words, he would suddenly raise his head, turn to the listeners, and during the entire forceful talk he would look intensely at them with his beautiful piercing blue eyes. It was unforgettable."

In astronomy, Gauss built a following that played a decisive role in the development of German astronomy. His circle of students included Schumacher, Gerling, Nicolai, Möbius (better known as a mathematician), Struve, Encke, and many others.

On the other hand one cannot say that Gauss built a school within mathematics. Riemann, Dedekind, and Eisenstein—who died too early to be able to redeem Gauss's great expectations of him—belonged, along with a few others, to that category to which Gauss alluded when he talked about the students who handle themselves best on their own and need only infrequent suggestions. Of these three it is thought that Riemann was the only one who became somewhat close to Gauss. Others, in themselves not insignificant mathematicians, did not experience quite the same relationship. Every time von Staudt gave Gauss the solution to a proposed problem, Gauss gave him his own in return, with the genial comment that he counted on mutual satisfaction. Gauss's humor was covered with frost from his Olympian mountaintop.

Mathematicians probably had little benefit from Gauss's perpetually repeated lectures in astronomy. But the situation would have been completely different had he continued along the lines indicated by the lectures on number theory and curved surfaces that were mentioned earlier. We ought to remember that the boundaries between different branches of science were not so clearly marked at that time, and besides that, due to his unparalleled position, Gauss could talk on any subject he wished. Thus we can imagine seminars on elliptic functions in the presence of Abel and Jacobi, or seminars on non-Euclidean geometry, first with Johann Bolyai—at one time Wolfgang wrote to Gauss and asked him to accept his son as a student, but he received no answer—and later with Riemann among the listeners around the table with the armchair and the little blackboard.

Alexander von Humboldt tried several times to talk Gauss into moving to Berlin as director of a research institute. His model was the *École polytechnique*, the technical university in Paris, which was founded in 1794. It was an important but often overlooked result of the French revolution. There, for the first time in the history of science, the principle was applied that all gifted persons, without respect to birth or fortune, should have an equal opportunity for instruction and that professors should take an active part in their education. The consequences of this principle for the development of the natural sciences and our present technical civilization can hardly be overestimated.

But the environment in which Gauss grew up was not yet influenced by such ideas, and Gauss's personal experiences in connection with Duke Ferdinand's death made him unfavorably disposed, for the most part, to the ideas of the French revolution. Nor did he ever forget his debt of gratitude to the Duke, who had paid his entrance to what was then the exclusive club of the university world. From this point of view we can look upon Gauss as one of the last—and greatest—cultural products of the feudal system of society.

He never accepted von Humboldt's invitation to move to the university in Berlin. Nor did he take an active part in training the young through direct influence as a teacher or through more personal contact. But this cannot be explained as solely a consequence of environment, not even to a substantial extent, since from his earliest childhood years Gauss was spiritually isolated from other people through his genetic superiority; perhaps this solitude was not solely a consequence of his genius, but rather an independent element of his character that existed from the very beginning.

We do not know much about Gauss's childhood other than those anecdotes that Gauss found it a pleasure to tell in his old age. But he gives the impression of having been mature from the very beginning, like Pallas Athena who sprang forth in full armor from the head of Zeus, the fount of all wisdom. He is thought to have played very early, not with material things as other children do but with numbers; and his own statement, that he could count before he could talk, certainly need not be an old man's romantic exaggeration.

Independent of his environment he exercised his gift for pure mathematics; he worked with unbelievable diligence and with an endurance which is supposed to never have failed. He calculated decimal expansions with close to a thousand decimals, which would seem almost completely improbable if we did not have his tables as undeniable evidence. He also learned how to handle series at an early age; in the beginning he used them to calculate square roots, logarithms, trigonometric functions, and so forth.

With enormous perseverance he built up his numerical virtuosity, and in his crystal-clear memory there remained for the rest of his life a panoramic view of the kingdom of numbers which is almost certainly unique in the history of mathematics.

But Gauss was not just a living book of tables to be looked into for necessary calculations. He experimented with his enormous wealth of numerical material and followed the path of induction to many of his most beautiful results, such as the "gem of arithmetic." After his creative intuition and numerical ability did their part, his logical rigor carried on to the arduous development of an incontrovertible proof.

It is this combination which is Gauss's unique characteristic among mathematicians. In addition to this, he by no means restricted himself to the theoretical level. Arithmetic was his favorite domain, yet he gladly dealt with the "few clods of soil that we call planets," whether it was a matter of determining the oribts of new asteroids or measuring the earth's magnetic field. He worked with equal ease in all of the areas of applied mathematics of his time. His ability in carrying out experiments and handling observed data is still a model for empirical research today.

Thus Gauss united several qualities, of which any one alone would have been adequate to produce a great scientist. All of these qualities—the creative intuition, the numerical virtuosity, the logical rigor, the experimental proficiency—were united in a harmonic ensemble for the solution of great or small problems over the entire field of natural science, and in the end the results he achieved poured forth in perfect form.

It used to be that only Archimedes and Newton were mentioned as his equals. Both possessed this improbable versatility: they commanded with equal ease all of the branches of the extant natural science and at the same time they carried them far beyond the bounds established by their predecessors. As a theoretician Einstein certainly falls into the same class, but in his case there is the limitation that he was not an empiricist. He made only mental experiments—which unquestionably revolutionized our world—but he left the empirical confirmation to others.

There is no reason to compare here the private lives of these geniuses, of which, in the case of Archimedes, we know little. We return to Gauss. It is an astonishing fact that he managed to live a normal, middle-class life, free from the blows that hard reality usually distributes so generously to persons who deviate too far from their surroundings. It seems as though his practical and experimental proficiency even helped him come to grips with the immediate reality of his everyday life.

Aside from isolated periods of depression caused by the deaths of friends or relations—or perhaps caused by undue strain due to a superhuman work load—he lived in harmony with his environment.

In spite of the Gaussian reticence there exists reason to believe that this harmony extracted its price. In Gauss's surroundings there was scarcely any person who—outside the sphere of everyday life—could share his thoughts or give him any new ones in return. When it came to the deepest ideas he had no one but himself to talk to. A journal note was made and other volumes had their entries. The feeling of loneliness must have grown with the years, and it is not to be wondered at that he developed an attitude that we have referred to several times earlier as one of Olympian loftiness and coldness. This bearing was also supported by a fame that only few scientists have ever come to enjoy. The feelings afforded him by those around him can only be characterized by words such as reverence and adoration.

This remoteness from other people did not only lead to an egotistical protection of his own working environment and to a judgment so objective that it often makes an unfavorable impression, but also to an alienation from the rest of the human community.

His solitude would probably have existed in whatever time and whatever environment Gauss had lived. But let us not for a moment believe that he ever really wanted to be a completely usual person, rather than the creator of a universal work with few counterparts in the history of mankind.

BIBLIOGRAPHY

The best source for Gauss and his work is still *Carl Friedrich Gauss' Werke*, Volumes I-XII, brought out by the Royal Scientific Society of Göttingen. The collection was edited by E. Schering, F. Klein, M. Brendel, and L. Schlesinger, with the assistance of R. Fricke, P. Stäckel, E. Wiechert, C. Schaefer, A. Galle, and H. Geppert. It was successively published by the publishing houses Perthes (Gotha), Teubner (Leipzig), and Springer (Berlin) during the years 1863-1933.

In it are Gauss's journal, part of his correspondence, and commentaries and essays about his scientific contributions to various fields. The following three important works of Gauss have been translated into English.

Gauss, Karl Friedrich, *Theory of the Motion of the Heavenly Bodies Moving About the Sun in Conic Sections*, translated and with an appendix by Charles Henry Davis. New York, Dover Publications, 1963.

Gauss, Karl Friedrich, *General Investigations of Curved Surfaces*, translated from the Latin and German by Adam Hiltebeitel and James Morehead. Introduction by Richard Courant. Hewlett, New York, Raven Press, 1965.

Gauss, Karl Friedrich, *Disquisitiones arithmeticae*, translated by Arthur A. Clarke. New Haven, Yale University Press, 1966.

A strongly concentrated and modern version of Gauss's work plus biographical data and commentaries is *C. F. Gauss 1777-1855. Gedenkband anlässlich des 100. Todestages am 23. Februar 1955*, put out by Hans Reichardt, with contributions by eleven experts. Teubner, Leipzig 1957.

The most complete biography of Gauss is *Carl Friedrich Gauss, Titan of Science*, by G. Waldo Dunnington, Exposition Press, New York, 1955. Among other things it contains a detailed list of the extensive literature about Gauss. This book was reissued in 1960 by Hafner Publishing Co, New York.

There is an essay on Gauss in *Men of Mathematics* by E. T. Bell, Simon and Schuster, New York, 1937. The same essay is included in the anthology *The World of Mathematics* by James R. Newman, Simon and Schuster, New York, 1956.

Gauss and his work are described in a simple and readable fashion in *Carl Friedrich Gauss, Prince of Mathematicians*, by William C. Schaaf. New York, Franklin Watts, Inc. 1964.

Sections concerning Gauss are to be found in the following three books:

Turnbull, H. W., *The Great Mathematicians*. Meuthen, London, 1929.

Scott, Joseph F., *A History of Mathematics from Antiquity to the Beginning of the 19th Century*. Taylor and Francis, London, 1958.

Muir, Jane, *Of Men and Numbers: The Story of Great Mathematicians*. Dodd, Mead, and Co, New York, 1961.

INDEX